Mohamed Chnafi
Omar Mommadi
Abdelaziz El Moussaouy

Semiconductor Nanostructures:

Mohamed Chnafi
Omar Mommadi
Abdelaziz El Moussaouy

Semiconductor Nanostructures:

Excitons and Sorting for Optoelectronic Applications

ScienciaScripts

Imprint

Any brand names and product names mentioned in this book are subject to trademark, brand or patent protection and are trademarks or registered trademarks of their respective holders. The use of brand names, product names, common names, trade names, product descriptions etc. even without a particular marking in this work is in no way to be construed to mean that such names may be regarded as unrestricted in respect of trademark and brand protection legislation and could thus be used by anyone.

Cover image: www.ingimage.com

This book is a translation from the original published under ISBN 978-620-6-72772-9.

Publisher:
Sciencia Scripts
is a trademark of
Dodo Books Indian Ocean Ltd. and OmniScriptum S.R.L publishing group

120 High Road, East Finchley, London, N2 9ED, United Kingdom
Str. Armeneasca 28/1, office 1, Chisinau MD-2012, Republic of Moldova, Europe
Printed at: see last page
ISBN: 978-620-4-74651-7

Table of contents

General introduction

Rapid progress in semiconductor technology in recent years has enabled the manufacture of small-scale electronic nanostructures. These nanostructures confine charged particles to three dimensions of space. In small dimensions, especially in quantum dots (BQs) [1,2], the picture is different because the nanostructures are less than a nanometer wide, a few nanometers thick, and in different shapes. In BQs, quantum confinement is important, leading to greater stability of excitons and trions by increasing their binding energies. The stability of such particles remains dependent on temperature and pressure. Adequate identification of the negative trion (X^-) was not determined until 1990 in high-quality, remotely-doped quantum wells [3-5]. Since then, extensive work has been carried out on (X^-) in two-dimensional quantum wells [6,7] and BQs, and the first observations of trions were made on an ensemble of BQs [8]. There are several theoretical studies dedicated to excitons [9-11] and trions [12-21] in this type of nanostructure. Most of these works have treated and considered spherical [22-24], disk-shaped [25,26], flat square [27,28], and cylindrical [29,30] QBs.

When optically exciting carriers in a semiconductor, the minimum energy required to form free carriers is the gap energy (the energy of the band gap). Energy below this value cannot excite free carriers. Indeed, the study of semiconductor absorption at low temperatures has shown excitation just below the gap [31]. This excitation is associated with the formation of a bonded electron and electron hole, also known as an exciton. This is an electrically neutral quasiparticle like the hydrogen state. At low temperatures, the bound states are formed and the Coulomb interaction between the electron and the hole becomes important [32]. The (X^-) is created due to the extra electron bound to a pre-existing exciton in the QB, and if a hole is bound to an exiton, a positive trion (X^+) is created. Positive and negative trions are complex electronic states excited in semiconductors and so the 3-body problem is raised. Although in 1958, Lampert [33] originally and theoretically predicted (X^-) in semiconductors, K.KHENG [34] experimentally resulted in (X^-) in the Cd Te/Cd Zn Te-based quantum well.

In this work, we have addressed the effect of quantum confinements (3D) on exciton binding energy and negative trion energies (X^-) under the combined influence of pressure and temperature, in a semiconducting cylindrical QB fabricated in$GaAs$ and surrounded by $GaAs/Ga_{1-x}Al_xAs$. Using the variational approach and the effective mass approximation considering finite and infinite confinement potential. There were concerns about the validity of the effective mass approximation in the BQ limit when the exciton size can be compared to average semiconductor lattice constants.

In the first chapter, we review the fundamental properties of semiconductors, as well as the models used to describe confined electronic states: effective mass approximation, envelope function, variational method, etc...

In the second chapter, we describe our theoretical model, which relies essentially on the vibrational method to study excitonic states in a cylindrical QB. We determine the exciton binding energies and the photoluminescence energy transition in different boundary cases generated by this geometry, with a view to showing the importance of the effect of temperature and hydrostatic pressure.

In the third chapter, we present the formalism of negatively charged excitons in a QB of cylindrical semiconductor geometry based on$GaAs/Ga_{1-x}Al_xAs$. We establish the Schrödinger equation for the envelope function in the effective mass approximation using a variational method, and discuss the invariants of the problem. Finally, we determine the energy of the negative trion subband in the case of the finite and infinite square well under temperature.

Bibliography

[1] Peter Ramvall, Satoru Tanaka, Philippe Riblet, and Yoshinobu Aoyagi, Appl.Phys.Lett. 73(1998)1104.

[2]L. Esaki, in physics and Application of Quantum wells and superlattice, vol 170, of NATO   Advanced study institute, series B: physics, edited by E.E, Mendez and K.vonklitzing (plenum,   New York, 1987).

[3]K. Kheng, R.T. Cox, Merle Y. dÁubigne, Franck Bassani, K. Saminadayar and Tatarennko, Phys.Rev.Lett. 71(1993)1752.

[4]G. Finkelstein, H. Shtrikman, and I. Bar-Joseph, Phys. Rev. Lett.74 (1995)976.

[5]A. J.Shields M. Pepper, D. A. Ritchie, M. Y. Simmons and G. A. C. Jones, Phys. Rev. B51 (1995)18049.

[6]D. Sanvitto, F. Pulizzi, A.J. Shields, P.C. Christianen, S. N. Holmes, M.Y. Simmons, D.A. Ritchie, J. C. Maan, and M. Pepper, Science 294.5543(2001) 837.

[7]G. Eyton, Y. Yayon, M. Rappaport, H. Shtrikman, and I. Bar-Joseph, Phys. Rev. Lett. 81(1998) 1666.

[8]Y. Yayon, M. Rappaport, V. Umansky and I. Bar-Joseph, Phys. Rev. B64 (2001) 81308.

[9]C. F. Lo and R. Sollie, Solid state commun 79 (1991)775.

[10]S. I. Pokutnil.phys. Semicond 25 (1991) 381.

[11]G. T.Einevoll, Phys. Rev. B45 (1992) 3410.

[12]M.A. Olshavsky, A. N. Goldstein, and A. P. Alivisatos, J. Am. Chem.Soc.112 (1990) 9438.

[13]W. Xie, and C. Chen, Physica E. 8 (2000) 77.

[14]W.F. Xie, Phys.Stat.Sol. B226 (2001)247.

[15]B.Stébe, A.Moradi, and F. Dujardin, Phys. Rev. B61 (2000) 7231.

[16]C.Riva F. M. Peeters, and K. Varga, Phys. Rev. B61 (2000) 13873.

[17]I. Szlufarska, A. Wojs, and J. J. Quinn, Phys. Rev. B63 (2001) 085305.

[18]W.Y.Ruan, K. S. Chan, and, E. Y. B. Pun, J. Phys.: Condens. Matter 12(2000) 7905.

[19]B.Szafran, B. Sté be, J. Adamowski, and S. Bednarek, J. Phys.: Condens. Matter 12 (2000) 2453.

[20]L. C. O. Dacal and J. A.Brum, Phys. Rev. B65 (2002)115324.

[21]I. M. Kupchak, Yu. V. Kryuchenko, and D. V. Korbutyak, Semiconductor Physics, Quantum Electronics & Optoelectronics, 9(2006) 1-8.

[22]J.L.Marn, R.Riera, and S.A.Cruz, J. Phys: condens. Matter 10(1998) 1349.

[23]Shudong Wu and Liwan, J. Appl. Phys.111 (2012) 063711.

[24]H.Hassanabadi and A.A.Rajabi, Phys. Lett. A373 (2009) 679.

[25]L.C.Lew Yan Voon and M. Willatzen, J. Phys: Condens. Matter 14(2002) 13667.

[26]S.Gaan, Guowei, and R. Feenstra, J. Appl. Phys 108 (2010) 114315.

[27]Szafran, B.Chwiej, T. Peeters, F. M. Bednarek, S. Adamowski, J. andPartoens, B. *Physical Review B, 71*(2005) 205316.

[28]Abbarchi, M., Kuroda, T., Mano, T., Sakoda, K., Mastrandrea, C. A., Vinattieri, & Tsuchiya, T. (2010). Energy renormalization of exciton complexes in GaAs quantum dots. *Physical Review B, 82*(20), 201301.

[29]Y.Kayanuma, Phys. Rev. B44 (1991) 13085.

[30]V.A. Holovatsky, M.J. Mikhalyova, and M.M. Tkach. J. Phys.: Condens. Matter. 3(2000) 863.

[31] Jacques I. Pankove, Optical Processes in Semiconductors, Prentice-Hall, Inc (1971).

[32]L.V.Keldy, P.N.Lebedev, Contemp.Phys.27 (1986)395.

[33] M.A. Lampert, Phys. Rev.Lett.1 (1958) 450.

[34]G. Finkelstein, V.Umansky, and I. Bar-Joseph, Phys. Rev. B58 (1998) 12637.

CHAPTER 1 : GENERAL INFORMATION ON SEMICONDUCTOR-BASED NANOSTRUCTURES

I- Introduction

Nanostructures are length-scale system materials in the range 1 to 100 nm in at least one dimension (1D). In a nanostructure, electrons are confined to the nanometric dimension, but are free to move in other dimensions. One way of classifying nanostructures is based on the dimensions in which electrons move freely:

- ❖ Quantum wells: electrons are confined to one dimension (1D), and remain free in the others. They can be created by sandwiching a narrow-bandgap semiconductor layer between the large gaps. A quantum well is often referred to as a 2D electron system [1-3].
- ❖ Quantum wires: electrons are confined in two dimensions, and remain free to move in the third dimension. Real quantum wires include polymer chains, nanowires and nanotubes [4].
- ❖ BQs: electrons are confined in all dimensions, as clusters and nano-crystallites [5]. Numerous studies are now focusing on semiconductor BQs capable of confining electrons on the nanometer scale in all spatial directions [6,7]

When a semiconductor material is structured on the nanometer scale, its electronic and optical properties are governed by quantum mechanics. The quantum well, formed by a thin semiconductor layer of nanometric thickness, has been widely used over the past 20 years to manufacture high-performance components (laser diodes, two-dimensional electron gas transistors, etc.) [8-12].

The field of nanostructures is becoming a very important one, and is undoubtedly at the heart of the next revolutions, thanks to the major applications expected in the life sciences, materials and information technologies.

II- Quantum box semiconductor

1- The three-dimensional crystal

Early work on semiconductor properties focused mainly on pure elements such as Silicon and Germanium, which are characterized by an indirect gap. These are elements in the 4th column of the periodic table. These elements have 4 electrons in their valence orbitals. Semiconductor materials can also be binary compounds of type II-

VI(CdSe, CdS, ZnSe, CdTe, etc ...)   ,   III-V   I-VII(GaAs, GaP, InP ...), (CuCl, CuBr, AgBr ...) or ternary compounds of type IIII-IV (CuInSe2, CuInS2...). In the case of binary compounds, each unit cell comprises two atoms with 8 valence electrons [13].

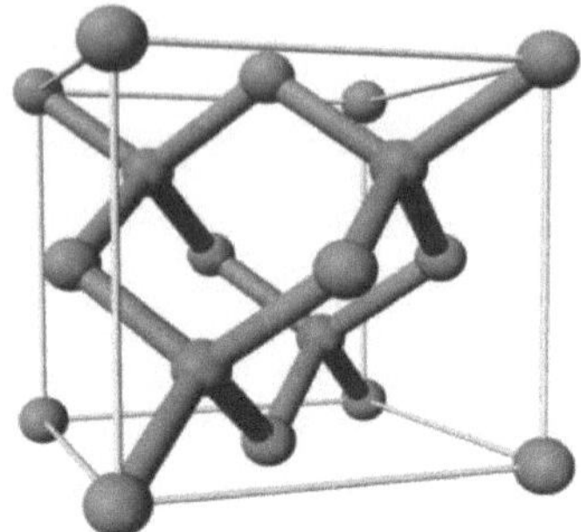

Figure 1.1: Zinc-blende crystal structure (Example of Gallium Arsenic (GaAs)).

In a three-dimensional semiconductor material, the density of states is a continuous function of energy. At 0 K, charge carriers occupy the valence band, which is separated from the conduction band by a zone known as the band gap, in which the Schrödinger equation admits no solutions. The height of the band gap is low, and conduction in this type of material increases with temperature.

In the late 60s, Leo Esaki [14] produced the first semiconductor heterostructure. This exotic artificial structure earned its inventor the Nobel Prize in Physics in 1973. Since then, theoretical and experimental studies on the subject of low-dimensionality have evolved from two-dimensional semiconductor structures such as quantum wells and superlattices to zero-dimensional structures, also known as BQs or quantum dots, and one-dimensional semiconductor structures, also known as quantum wires.

As we've just seen, we can distinguish between different types of semiconductor nanostructures, depending on the degree of freedom of the charge carriers in the structure and on the geometry.

2- Two-dimensional (2D) structures

These structures allow charge carriers freedom of movement in two directions in space. Conduction-band electrons and valence-band holes remain confined within the smaller-gap material, which acts as a potential well. In the direction normal to the wells, energies are quantized.

Structures (2D) are classified into three categories:

➤ Simple quantum wells :

A simple quantum well consists of a semiconductor layer whose thickness is of the order of magnitude of the De Broglie wavelength of the electrons, and whose other dimensions are very large compared to this wavelength [17].

➤ Multiple quantum wells:

A multiple quantum well is made up of a juxtaposition of single wells, close enough for there to be coupling between the electron motions in the different wells [17].

➤ Super networks :

A superlattice is a multiple quantum well with a large number of equidistant single wells, so that the structure can be considered quasi-periodic in the direction normal to the wells. **Figure 1.2** shows the first structure proposed by Esaki and Tsu [14,18]. It consisted of an alternating stack of thin layers of two semiconductors [18].

Figure 1.2: Leo ESAKI, winner of the 1973 Nobel Prize in Physics. Diagram of a superlattice.

3- One-dimensional structures (1D)

In these structures, confinement affects two dimensions, leaving charge carriers free to move in only one direction (1D). The result is quantum wires (Qw). The density of states, which is staircase-like in the case of quantum well structures, then features peaks that are all the finer the smaller the cross-section of the quantum wire.

4- Quantum boxes (0D)

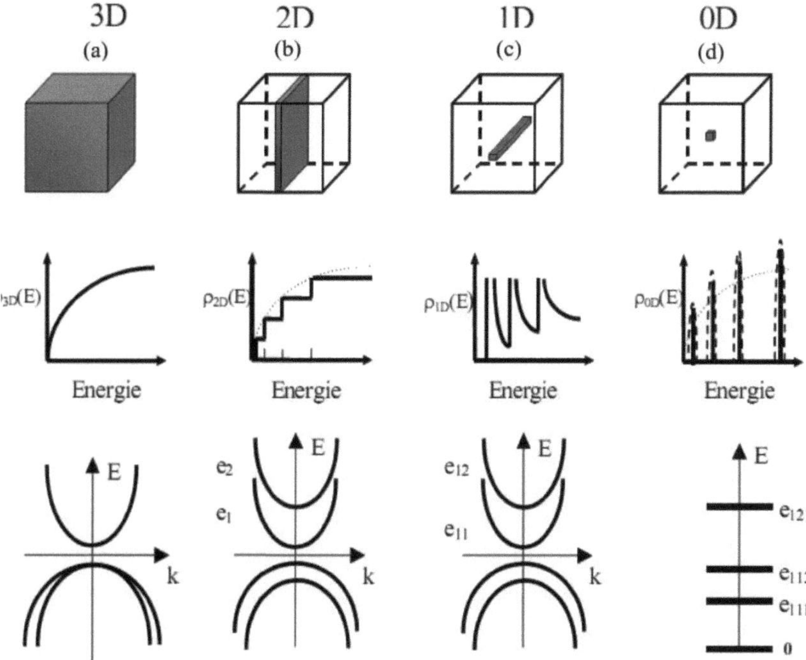

Figure 1.3: *Evolution of density of states ρ(E) and band structure E(k) as a function of material dimensionality. (a) a bulk semiconductor (3D), (b) a quantum well (2D), (c) a quantum wire (1D) and (d) a quantum box (0D).*

In the 1980s, technological advances in the growth of heterostructures led to the emergence of a new type of three-dimensional nanoscale structures known as quantum dots or BQs (Quantum Boxes or Quantum Dots). The term Quantum Dot was coined by Mark Reed. These artificial structures are also known as "artificial atoms". The origin of this nickname comes from the discretization of the energy spectrum induced by the three-dimensional confinement of charge carriers in these systems.

Depending on the manufacturing process, semiconductor BQs are either embedded in crystalline or semiconductor matrices [19] or embedded aqueous or organic solutions [20,21]. Charge carriers are deprived of their freedom of movement all spatial directions.

As a result, they are confined a region whose dimensions are of the order of the De Broglie electron wavelength. This confinement gives BQs intermediate properties between the classical bulk semiconductor and the isolated atom.

III- Quantum box manufacturing :

1- Historical overview :

Semiconductor nanocrystals have been around for a very long time. In ancient times, the Egyptians made hair dyes from lead salts. On contact with the scalp, the semi-conductingPbS nano-crystals that formed gave the hair a black color [22]. In the early 1980sAlexei Ekimov's team [23,24] synthesized glass doped with semiconductor nanocrystals, A. Henglein's group [25] developed protocols for preparing semiconductor colloids in solution. The development of the qualities of these nanostructures and the study of their size-dependent properties is the responsibility of A. Ekimov nanocrystals in glass matrix [23-26] and L. Brus for colloidal nanocrystals in solution [21-27]. The improvement and study of the theory associated with these nanostructures was rapidly undertaken by Al. Efros [28]. These early studies marked the emergence of a new field of research into synthesis and physical properties of colloidal semiconductor nanocrystals.

In 1993, the first organometallic synthesis of semiconductor nanocrystals was carried out by C. Murray, et al [29]. These studies resulted in dimensionally controlled nanocrystals with low size dispersion, high crystallinity and efficient surface passivation.

2- Molecular-jet epitaxy :

The principle of molecular beam epitaxy (MBE) was first developed at the turn of the century, but it wasn't until the late 1960s that it was perfected [30]. , this technique is widely used for the growth of thin films, and in particular for the growth of semiconducting BQs, as it abrupt and crystallographically perfect interfaces to be obtained. EJM is a vacuum deposition technique (residual pressure between 10^{-10} et 10^{-9} Pa) to ensure surface cleanliness, as without a vacuum the atoms used for growth can be mixed with residual gas molecules. The principle consists in evaporating materials from elementary sources in effusion cells in the form of molecular jets, and directing these material flows towards a heated crystalline support (substrate). The decomposed atoms diffuse onto the substrate surface to form monolayer by monolayer stacks of thin films. (J. Zribi, 2008).

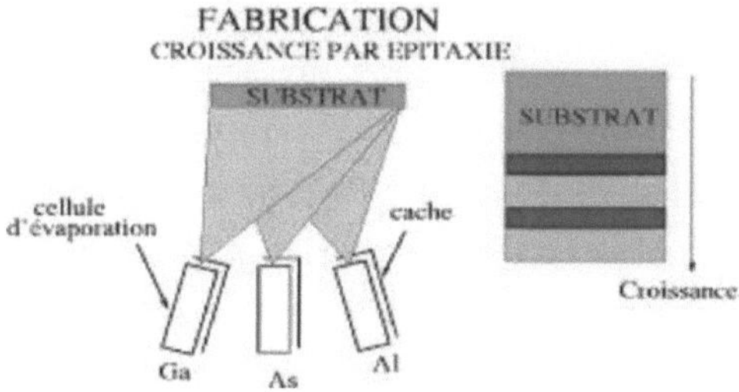

Figure 1.4: Schematic diagram of molecular beam epitaxy growth.

3- Quantum box growth methods

The growth of BQs has always been of great scientific, applied and, above all, technological interest. It has undergone a major evolution thanks to the development of methods such as molecular jet epitaxy, which have enabled the deposition of very thin layers on solid substrates. The leitmotiv of researchers working in this field has been: the better we understand the fundamental processes involved in growth, the better the quality and performance of devices based on semiconductor nanostructures.

The first semiconductor BQs were manufactured using a lithography and etching technique. Unfortunately, this method resulted in BQs with several surface defects and small sizes. In 1985, Goldstein and his team were able to produce the first BQs by epitaxial growth [31].

IV- Quantum box applications

Today, the study of BQs is proving extremely important, as it will enable the scientific community to develop the physics and behavior of these mesoscopic systems, with a view to improving knowledge, advancing the state of the art of matter, acquiring technological mastery on the nanometric scale, and finally seeking new functionalities in an exciting field straddling condensed matter physics and quantum physics. Nanoscale structures offer novel and highly interesting electrical, magnetic, optical and chemical properties that cannot be found in bulk semiconductors.

1- Applications in optoelectronics and photonics

Semiconductor nanostructures are currently of great interest. Thanks to their exceptional electronic and optical properties, and to technological advances in manufacturing and synthesis methods, these structures can be easily integrated into numerous optoelectronic devices, mainly in the field of optical telecommunications, such as laser diodes, optical amplifiers, saturable absorbers and photodetectors. They can also be used to make photovoltaic cells and light-emitting diodes. The specific properties of BQs (discrete levels, fine emission lines and high radiative efficiency close to 100%) also make it possible to design innovative optoelectronic components based on a single BQ as a single-photon source. BQs have even supplanted other heterostructures such as quantum wells or quantum wires.

2- Applications in biology

In the field of biology, the teams of P. Alivisatos et al [32] and S. Nie et al [33] were the first to consider the use of fluorescent semiconductor nanocrystals as probes for cell labeling. These nanocrystals are inorganic fluorophores that are brighter than organic fluorophores. They can replace them in a number of applications, as they are resistive to photobleaching and can visualize several colors simultaneously.

The principle behind this detection technique is based on the fact that these nanostructures are soluble in aqueous media. Once in living tissue, these structures can attach themselves to biological molecules such as proteins or nucleic acids (DNA and RNA). The properties of these molecules can then analyzed by detecting the fluorescence of nanocrystals after they have been excited either by a laser or by a microscope's UV lamp. Nanocrystals play an important role biological analysis and medical diagnostics.

The use of BQs in biology, biotechnology, biomedical imaging (optical detection of certain cancerous tumors) or medicine is of interest to the scientific community in three different disciplines: physics, chemistry and medicine. However, a great deal of work remains to be done to develop, optimize and control these applications, as research in this field is still in its infancy.

However, research in this field is very active. It's easy to imagine that, in addition to the above-mentioned exploitations, other unexpected ones, taking advantage of the astonishing characteristics of these nano-objects, should see the light of day in the future.

V- Band properties

The energy of the band gap, noted Eg , is a material constant that depends only on temperature. The width of this energy band defines the material (semiconductor, insulator, conductor). For semiconductors, the gaps are of the order of magnitude of 1eV. Insulators are characterized by large gaps, typically Eg >5eV at room temperature. Metals are characterized by a very small gap, almost = 0 eV (no gap).

There are two types of gap:

- **Direct gap:** if the minimum of the upper band corresponds to the same wave vector

  as the maximum of the lower band.

- **Indirect gap:** the valence band maximum is not opposite the conduction band minimum.

  conduction band (they correspond to different k wave vectors).

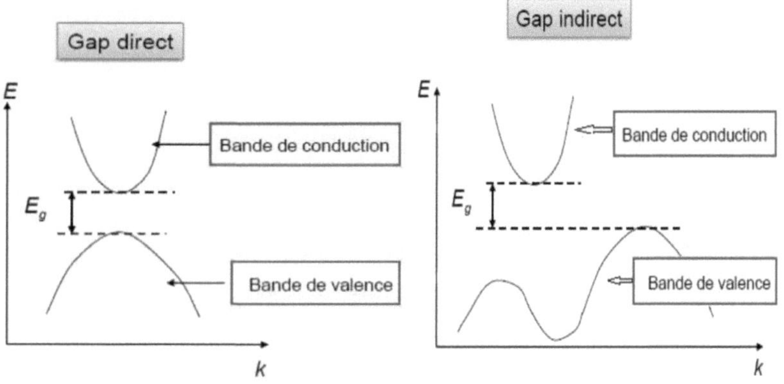

Figure 1.5: Diagram of a direct and indirect gap.

A few examples of prohibited strips

At T=300 K, we have the following gaps:

- For a semiconductor, silicon (Si) is characterized by a bandgap of $E_g(Si) = 1.12$ eV.
- For an Arsenic Gallium semiconductor of (GaAs) has a bandgap energy of bandgap energy of $E_g(GaAs) = 1.42$ eV.
- The germanium (Ge) semiconductor is characterized by a bandgap energy of energy of $E_g(Ge) = 0.66$ eV.

VI- Essential concepts and approximations

1- Hartree-Fock approximation

The Hartree-Fock approximation replaces the interaction of each electron in the atom with all the others with the interaction with an average field created by the nuclei and all the other electrons. In other words, the electron moves independently in an average field created by the other electrons and nuclei. This approximation reduces the N-body problem to a single-electron problem.

2- Born-Oppenheimer approximation

According to Born-Oppenheimer (1927), the only way to simplify the complication of solving the stationary Schrödinger equation is to study the motion of electrons and nuclei separately. In this case, the system Hamiltonian will be divided into two parts, one electronic and the other nuclear. This approximation is based on the adiabatic Born-Oppenheimer approximation, which relies on the large mass difference between electrons and nuclei [34].

Remember that nuclei are very heavy compared to electrons (around 1836 times), means that electrons can move faster in the crystal than nuclei, so the motion of the latter is negligible. Their kinetic energy is then assumed to be zero, and the potential energy of interaction between the nuclei becomes constant [35]. This approach leads to a Hamiltonian describing electron motion in a field created by a static configuration of nuclei [36].

3- Approximation of effective mass

In a classical description of effective mass, only the interactions between the electron and its environment are taken into account. In terms of this description, a particle evolving in a finite-dimensional crystal is subject to two forces: an external force Fi and another **R** representing the resultant of all the interaction forces between the particle and the crystal. In classical mechanics, the mass m_0 of the particle is described by the fundamental principle of dynamics:

$$R + Fi = \frac{dp}{dt}, \tag{1.1}$$

with p=m_0 v where p is the particle's momentum since mass is constant (not time-dependent), so equation (1) becomes :

$$R + F_i = m_0\gamma .\qquad(1.2)$$

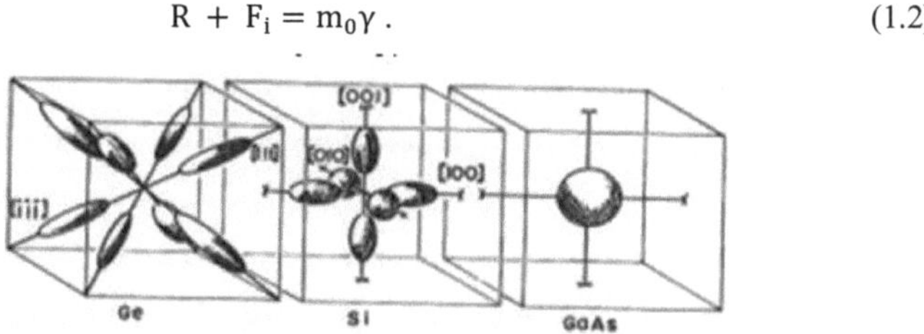

Figure 1. 6*: E(k) = constant surfaces for the conduction band of Ge, Si and GaAs in the effective mass approximation.*

We introduce a fictitious particle of massm^* (only the external force intervenes) whose state remains identical to that of the real particle (acceleration, energy). In a crystal, the electron's mass is defined by :

$$\left(\frac{1}{m^*}\right)_{i,j} = \frac{1}{\hbar^2}\frac{\partial^2 E}{\partial k_i\,\partial k_j}.\qquad(1.3)$$

For a three-dimensional crystal, the result is no longer trivial and takes the form of a nine-component tensor [37]. It is denoted m^* , and the isotropic case means E(k) has parabolic behavior. The constant-energy surfaces are ellipsoids centered on the zone extrema. The effective mass is isotropic for direct-gap semiconductors, while in the case of indirect-gap semiconductors we can define two components, one longitudinal and the other transverse in all spatial directions.

If a single electron is missing, the resulting conduction of the N-1 electrons is equivalent to that of a single positive particle (hole) with a speed equal to that of the missing electron. This is not the case for a completely filled valence band (no transport). **Table 1.1** shows the different values of effective charge carrier masses and gap energies inGaAs et CdTe materials, generally obtained by cyclotron resonance.

materials	E_g(eV)	$m_e()m_0$	$m_{lh}()m_0$	$m_{hh}()m_0$	Ref.
GaAs	1.424	0.0665	0.094	0.34	[38,39]
CdTe	1.606	0.098	0.11	0.662	[40,41]

Table 1.1*:Gap energy and effective masses of charge carriers (electron and hole) in selected hole) in some materials.*

The effective mass approximation (one-band method $\vec{k}.\vec{p}$) allows interactions between different energy bands to be taken into account to first order. It therefore adds a perturbation to the effective Hartree-Fock potential, whatever its origin (electric field, magnetic field, material heterogeneity). This formalism essentially provides qualitative information, and the bands are only parabolic in the vicinity of the center of the Brillouin zone (k=0). The validity of this model is thus limited to a fraction of the Brillouin zone (a few percent) in certain materials such as InAs ou InSb [42]. One of the main advantages of this theory is that it can be adapted to a wide range of physical situations.

4- Approximation of the effective mass for a non-degenerate band

The one-electron Hartree Fock equation can be written in the Born Oppenheimer approximation [34] as :

$$H_{HF}\varphi_{\lambda i}(\xi) = \epsilon_i\varphi_{\lambda i}(\xi) \tag{1.4}$$

where H_{HF} is written as the sum of the operator describing electron kinetics and the effective potential containing electron-ion, electron-electron and electron-ion exchange Columbian interactions, under the conditions imposed by lattice periodicity.

$$H_{HF} = -\frac{\hbar^2}{2m}\Delta + V_{eff}. \tag{1.5}$$

In equation (1.4), the λ_i represent the band indices, with spin effects neglected, the $\varphi_{n,k}(r)$ functions respecting Bloch's theorem, i.e. :

$$\varphi_{n,K}(r) = \exp(ikr)U_{n,K}(r). \tag{1.6}$$

The periodicity of the network allows us to write :

$$U_{n,K}(r + R_L) = U_{n,K}(r) \tag{1.7}$$

where R_L is the direct network vector.

Taking into account the spin-orbit interaction [43], the Hamiltonian is transformed as follows:

$$H_{eff} = -\frac{\hbar^2}{2m_e^*}\Delta + V_{eff}(r) + \left(\frac{1}{2m_e^*c}\right)^2 \sigma \wedge gradV_{eff}(r).p, \tag{1.8}$$

with σ the spin operator and c the celerity of light.

Adding a perturbation term $U(r)$ to the effective Hartree Fock equation leads to an equation of the form :

$$\{H_{HF} + U(r)\}\psi(r) = E\psi(r). \tag{1.9}$$

The eigenfunction ψ of the perturbed Hamiltonian can be developed in a basis consisting of the eigenfunctions and eigenvalues of H_{HF} :

$$\psi(r) = \sum_{n,k} \chi_n(k)\varphi_{n,k}(r). \tag{1.10}$$

Using the expression of $\psi(r)$ in equation (1.8), we obtain the following expression:

$$(\varepsilon_n(k) - E)\chi_n(k) + \sum_{n',k'} < \varphi_{n,k}(r)|U(r)| \varphi_{n',k'}(r) > \chi_{n'}(k') = 0. \tag{1.11}$$

The matrix element of the disturbance potential can be written as :

$$< \varphi_{n,k}(r)|U(r)|\varphi_{n',k'}(r) >= \int \exp(i(k - k')r)\, U^*_{n,k}(r)U_{n',k'}(r)U(r)\, d^3r, \tag{1.12}$$

the term $U^*_{n,k}U_{n',k'}$ can be developed as a Fourier series:

$$U^*_{n,k}(r)U_{n',k'}(r) = \frac{1}{(2\pi)^3}\sum_{k} C(nl, n'k', K)\exp(ikr), \tag{1.13}$$

where K is a vector of the reciprocal network. Equation (1.11) can be written as follows:

$$< \varphi_{n,k}(r)|U(r)|\varphi_{n',k'}(r) >= \sum_{k} \bar{U}(k - k' + K)C(nl, n'k', K). \tag{1.14}$$

The slow variation of the U(r) function implies a Fourrier transform $\bar{U}(K)$ containing only low spatial frequency terms. As a result, the study is limited to states near the top of the valence band or the bottom of the conduction band, located at $K = 0$ (direct gap). The matrix elements of the potential then become:

$$< \varphi_{n,k}(r)|U(r)|\varphi_{n',k'}(r) >= \bar{U}(k - k')C(nk, n'k', 0). \tag{1.15}$$

Following the development in a basis consisting of the functions and eigenvalues of H_{HF} with the perturbation theory $k.p$ [43], the approximate expressions of the functions $U_{n,k}(r)$ and the energy En(k) are written as follows:

$$U_{n,k}(r) = U_{n,0}(0) + \frac{\hbar}{m}\sum_{p \neq n} \frac{Kp_{pn}}{E_n(0) - E_p(0)} U_{p,0}(r), \tag{1.16}$$

$$E_n(K) = E_n(0) + \frac{\hbar^2}{2m}\sum_{\mu\nu} \frac{m}{m^*_{\mu\nu}} K_\mu K_\nu , \tag{1.17}$$

with μ, ν corresponding to the x, y, z directions and $m^*_{\mu\nu}$ is the effective mass tensor written as :

$$\frac{m}{m^*_{\mu\nu}} = \delta_{\mu\nu} + \frac{2}{m}\sum_{p \neq n} \frac{p^\mu_{np} p^\nu_{pn}}{\varepsilon_n(0) - \varepsilon_p(o)}. \tag{1.18}$$

Where p^μ_{pn} represents the matrix element of the p^μ component of the pulse:

$$p^\mu_{pn} =< \varphi_{n,0}(r)|p_\mu|\varphi_{p,0}(r) >. \tag{1.19}$$

Since we are in the case of a non-degenerate band, only the diagonal terms are non-zero and the dispersion relation (1.16) becomes :

$$E_n(K) = E_n(0) + \frac{\hbar^2}{2m}\Sigma_{\mu\nu}\, K^2. \tag{1.20}$$

F(r) being the Fourier transform of $\chi_{n(k)}$, it is a solution of the effective mass equation described in impurity theory by the formula :

$$\left[-\frac{\hbar^2}{2m}\Delta + U(r)\right]F(r) = [E - \varepsilon_n(0)]F(r). \tag{1.21}$$

With the aid of a few transformations, we find the effective mass equation usual in impurity theory, in the form of a relationship between the wave function $\psi(r)$ and the envelope function :

$$\psi(r) = F(r)U_{n,0}(r) + \frac{1}{m}\Sigma_{p\neq n}\frac{-i\nabla F(r)p_{pn}}{\varepsilon_n(o)-\varepsilon_p(0)}U_{p,0}(r).$$

(1.22)

5- Variational method

The main work devoted to the study of excitons in QBs is due to Bryant (Bryant 1987, 1988) [44,45]. In the case of an infinite confinement potential, he used a variational method to determine the ground state of the $\hat{H}$ operator of confined excitons. The result is very simple but very useful in many quantum physics problems, where it is impossible to know the exact solution.

The fundamental energy of the $\hat{H}$ operator is given by :

$$E_0 = \min_{\phi\in\mathcal{H},\phi\neq 0}\frac{\langle\phi|\hat{H}|\phi\rangle}{\langle\phi|\phi\rangle}, \tag{1.23}$$

i.e. E_0 is the minimum of the average energy obtained over all possible quantum states.

If the spectrum of $\hat{H}$ is : $\hat{H}|\psi_n\rangle = E_n|\psi_n\rangle$, we have $|\phi\rangle = \Sigma_n|\psi_n\rangle\langle\psi_n|\phi\rangle$ and $E_{(0)} \leq E_n$,

Therefore: $\langle\phi|\hat{H}|\phi\rangle = \Sigma_n E_n|\langle\phi|\psi_n\rangle|^2 \geq E_0\,\Sigma_n|\langle\phi|\psi_n\rangle|^2 = E_0\langle\phi|\phi\rangle$, (1.24)

So $E_0 \leq \frac{\langle\phi|\hat{H}|\phi\rangle}{\langle\phi|\phi\rangle}$. In the case of the ground state $\phi = \psi_0$, we have equality.

We choose a family of quantum states $|\phi_{\alpha,\gamma}\rangle$ called the test family, depending on the parameters . α, γ

We calculate $E_{\alpha,\gamma} = \dfrac{\langle \phi|\widehat{H}|\phi\rangle}{\langle \phi|\phi\rangle}$ and look for the α, γ values that make $E_{\alpha,\gamma}$ minimal.

VII- Effects of confinement

Charge carriers can be associated with their de Broglie length, given by the expression :

$$\lambda_{deB} = \frac{h}{\sqrt{2\pi m^* k_B T}} \ .$$

(1.25)

In semiconductors, de Broglie wavelengths for electrons at the temperature of liquid helium (4 K) are typically tens of nanometers. If one or more dimensions of the structure are comparable to this value, spatial confinement occurs, altering the energy levels and density of electronic states. In bulk semiconductors, the density of states is proportional to $\sqrt{E}$. In a quantum well, carriers are confined in one direction, and the density of states is independent of energy for each confined band. In a quantum wire, carriers are confined in two spatial directions, and the density of states is proportional to $\frac{1}{\sqrt{E}}$. Finally, in the case of confinement along all three spatial directions, the energy levels are completely discretized. *Figure 1.7* shows the density of electronic levels for 3D (bulk semiconductor), 2D (quantum well), 1D (quantum wire) and 0D (quantum box) structures.

Band structures for BQs can be represented by one of the following models: a cubic potential (confinement by finite or infinite barriers along the three axes of space), a spherical barrier, or by a harmonic potential. A commonly used approach is to approximate confinement along the Oz growth axis with a quantum well and consider the first confined state; then, confinement in the xOy plane is modeled by a parabolic potential. The solutions for the energy levels are named similarly to the energy levels of the electron in an atom. The result of the calculation of eigenstates and eigenenergies depends on the model chosen, but the introduction of 3-dimensional spatial confinement always has the following consequences: the density of states takes the form of a series of Dirac peaks (discretization of the density of states), the effective gap increases, and the oscillator force associated with radiative recombination increases.

In practice, the creation of confined structures involves growing semiconductor layers with different gap energies. This makes it possible to create heterostructures in which variations in band structure (conduction and valence) lead to the formation of confinement barriers. The size of the box and the height of the barriers define the number of states confined within it. ***Figure 1.8*** shows the lateral cross-section of a QBGaAs/AlGaAs made fromAlGaAs and GaAs layers. The offset energies for the valence and conduction bands are 28% and 72% respectively [46].

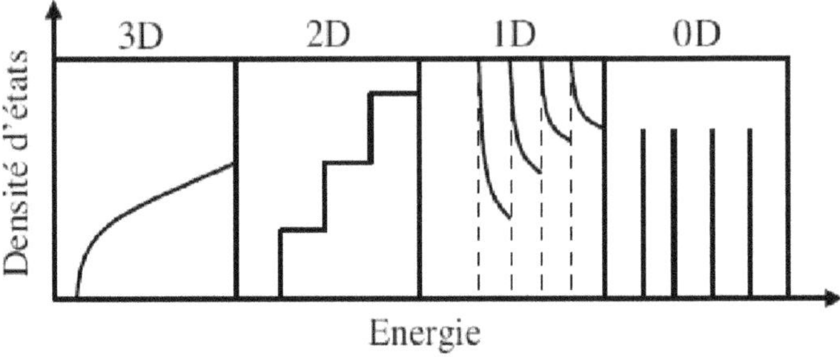

***Figure 1.7**: Density of electronic states in confined structures.*

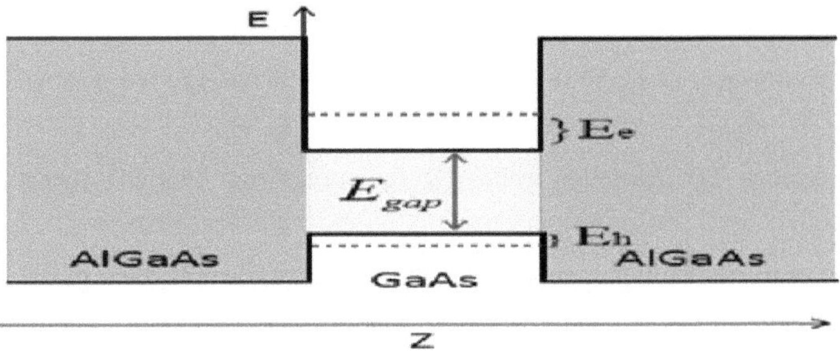

***Figure1.8**: band structure of a quantum box . GaAs/AlGaAs*

The transition energies can then be presented as the sum $E_e + E_h + E_g$ orE_e andE_h are proper energies in the conduction band and valence band,E_g is the gap energy. Note also that the shape of the well is different for boxes GaAs /GaAlAs , and InAs/GaAs . Stresses which are present in InAs/GaAs structures, and negligible in GaAs /GaAlAs systems, distort the potential, and the final band structure has a shape similar to that calculated in Ref. [47].

In the absence of strain effects, both valence bands (LH and HH) are degenerate at theΓ point in bulk semiconductors such as GaAs . Quantum confinement lifts the degeneracy between the heavy and light hole states, and the band energy of light holes decreases relative to that of heavy holes.

1- Effect of quantum box size

BQ size is one of the most important parameters to control in order to obtain sufficiently precise electronic properties for the various applications. Their size must be smaller than the thermal De Broglie wavelength to ensure separation of BQ energy levels.

2- Effect of pressure and temperature

V-element pressure has a major influence on crystal quality, especially at low temperatures. An increase in the amount of V-element present on the surface reduces the mobility of the various III-elements. In addition, at low temperatures, there is excess incorporation of arsenic into the matrix. These two phenomena can lead to an increase in surface roughness.

Lattice temperature is also a factor in the emission spectrum of cans. There is a red shift in the emission line, which essentially follows the temperature behavior of the bandgap [48].

In some cases, it is possible to observe a correlation between the red shift (for T < 100 K) of the emission line and a reduction in the inhomogeneous line width [49]. These observations reflect a transfer of charge to larger boxes (with lower energy levels). At higher temperatures, the line shift becomes more pronounced and the line width increases as carrier-phonon scattering and carrier thermal distribution become important [50].

By definition$E_g(P, T)$ is the pressure- and temperature-dependent band gapGaAs at the center Γ of the Brillouin zone. It can be expressed as [51, 52]:

$$E_g(P, T) = 1.519 - \frac{5.4 \times 10^{-4} T^2}{T + 204} + 0.01261\,P + 3.77 \times 10^{-5} P^2. \qquad (1.26)$$

3- Alloys

It is therefore possible to form a so-called ternary or quaternary solid by mixing two or three III-V semiconductors. However, the structure of this alloy is not that of a perfect crystal, due to the random distribution of atoms at each site of the zinc-blende structure, which in particular precludes the property of invariance by translation.

In order to describe the electronic states of the alloy, the virtual crystal approximation is often used. In such a model, the aperiodic potential is replaced by an average. For example, if we consider a solid $AB_{1-x}C_x$, atom A occupies the sites of the first face-centered cubic (FCC) sublattice of the zinc-blende structure, and atoms B and C randomly occupy the sites of the second FCC sublattice. The random potential created by B(VB) and C(VC) will be replaced by a periodic potential whose value is given by a linear interpolation between VB and VC :

$$\langle V \rangle = V_A + x.V_c + (1 - x).) V_B, \tag{1.27}$$

where x is the concentration of C in the crystal.

Translational invariance is restored, as are the symmetry properties of the zinc blende structure.

4- Photoluminescence

Photoluminescence is a process of photon emission caused by light excitation. In semiconductors, this photoluminescence originates from the radiative relaxation of a conduction electron to an empty valence-band state. This mechanism is also interpreted as the radiative recombination of an electron-hole pair, which emits a photon whose energy corresponds to the transition energy. This energy is greater than the gap energy of the BQ semiconductor, and provides access to confinement energies. This experimental technique makes it possible to study the electronic properties of semiconductor nanostructures. It requires electrons to be excited into the conduction band using suitable light excitation.

Conclusion

In this chapter, we provide an introduction to nanostructures and their properties, followed by a brief overview of the best-known fabrication methods. This is followed by a reminder of the methods used to study nanostructures, namely Hartree fock theory, the Born-Oppenheimer approximation, the variational principle, and the theory of effective mass at a non-degenerate band. Each of these methods is adapted to the specific situation, depending on the physical properties to be calculated. However, the variational method remains a simple one, and one that has shown its merits by judging the volume of published work using this theory.

To study the properties of excitons and trions in BQs, we have mainly focused on cylindrical box geometries. This choice is based on the fact that this geometry is more attractive for applications than others.

Bibliography

[1]Dallesasse, J. M., Holonyak Jr, N., Sugg, A. R., Richard, T. A., and El-Zein, N. *Applied Physics Letters*, 57(1990)2844-2846.

[2]Qian, F., Li, Y., Gradečak, S., Park, H. G., Dong, Y., Ding, Y.,Lieber, C. M. *Nature materials*, 7(2008) 701.

[3]Tiutiunnyk, A., Duque, C. A.a and Mora-Ramos, M. E. *Optical and Quantum Electronics*, 50(2018)234.

[4]J. E. Davey and T. Pankey,J. *Appl. Phys.* 39 (1968) 1941.

[5]A. Y. Cho, *J. Appl. Phys.* 42 (1971) 2074.

[6]El Hadi, El Moussaouy, Nougaoui, Bria, *jmaterenvironsci*, 8 (2017) 911-920.

[7]A. El Moussaouy, D. Bria, A. Nougaoui, *SolarEnergyMaterials&SolarCells90* (2006) 1403-1412.

[8]L. Pfeiffer, Wand K. W., Stormer H, L. *Appl. Phys. Lett.* 56 (1990) 1697-1699.

[9]O. Ambacher, J. Majewski, C. Miskys, A. Link, M. Hermann, M. Eickhoff,M. Stutzmann, F. Bernardini, V. Fiorentini, V. Tilak, B. Schaff, and L. F. Eastman,*Journal ofPhysics: Condensed Matter*, 14 (2002) 3399.

[10]G. Koley and M. G. Spencer, *Applied Physics Letters*, vol. 86 (2005) 042107.

[11]D. Dumka, C. Lee, H. Tserng, P. Saunier, and M. Kumar, *Electronics Letters*, 40(2004) 1023-1024.

[12]J. Johnson, E. Piner, A. Vescan, R. Therrien, P. Rajagopal, J. Roberts, J. Brown,S. Singhal, and K. Linthicum, *Electron Device Letters, IEEE*, 25 (2004) 459-461.

[13]Javaux, C. Étude de la réduction du phénomène de clignotement dans les nanocristaux semi-conducteurs de CdSe/CdS à coque épaisse (Doctoral dissertation, Université Pierre et Marie Curie-Paris VI).(2012).

[14]Esaki, L., and Tsu, R. *IBM Journal of Research and Development*, 14(1970) 61-65.

[15]Bastard, Gerald. "Wavemechanicsapplied to semiconductorheterostructures." (1990).

[16]Assaid, E. M. (1995). Etude des propriétés électriques et optiques des complexes (D+, X) dans les microcristallites de semi-conducteur de forme sphérique (Doctoral dissertation, Metz).

[17]Le Goff, S., and Stébé, B. *Journal of Physics B: Atomic, Molecular and Optical Physics*, 25(1992) 5261.

[18]Tsu, R., and Esaki, L. *Applied Physics Letters*, 22(1973) 562-564.

[19] Ekimov, A. I., andOnushchenko, A. A. *Jetp Lett*, 34(1981) 345-349.

[20]Brus, L. E. *The Journal of chemical physics*, 79(1983)5566-5571.

[21]Brus, L. E. *The Journal of chemical physics*, 80(1984) 4403-4409.

[22] Walter, P., Welcomme, E., Hallégot, P., Zaluzec, N. J., Deeb, C., Castaing, J., andTsoucaris, G. *Nano letters*, 6(2006) 2215-2219.

[23]Ekimov, A. I., andOnushchenko, A. A. *Jetp Lett*, 40(1984) 1136-1139.

[24]Ekimov, A. I., &Onushchenko, A. A. *Soviet Physics Semiconductors-Ussr*, 16(1982)775-778.

[25]Alfassi, Z., Bahnemann, D., &Henglein, A. *The Journal of Physical Chemistry*, 86(1982)4656-4657.

[26] Ekimov, A. I. (1985). AI Ekimov, Al. L. Efros and AA Onushchenko, *Solid State Commun.* 56, (1985) 921-924.

[27]Brus, L. E. Journal of Luminescence, 31 (1984)381-384.

[28]Efros, A. L., andEfros, A. L. *Soviet Physics Semiconductors-Ussr*, 16(1982) 772-775.

[29]Murray, C., Norris, D. J., andBawendi, M. G. *Journal of the American Chemical Society*, 115 (1993)8706-8715.

[30]Jensen, P. *La Recherche*, 283(1996)42-47.

[31]Goldstein, L., Glas, F., Marzin, J. Y., Charasse, M. N., and Le Roux, G. *Applied Physics Letters*, 47 (1985)1099-1101.

[32]Peng, X., Wickham, J., andAlivisatos, A. P. *Journal of the American Chemical Society*, 120(1998)5343-5344.

[33]Smith, A. M., andNie, S. *Accounts of chemical research*, 43(2009)190-200.

[34]Born, M., & Oppenheimer, R. *Annalen der physik*, 389(1927) 457-484.

[35]Parr, R. G. W. Yang. "Density-functional theory of atoms and molecules." (1989).

[36]Kohn, W. *Reviews of Modern Physics*, 71(1999)1253.

[37]Chelikowsky, J. R., & Cohen, M. L. *Physical Review B*, 14(1976) 556.

[38]Le Goff, S., and Stébé, B. *Physical Review B*, 47(1993)1383.

[39]Susa, N. *IEEE journal of quantum electronics*, 32(1996) 1760-1766.

[40]Deleporte, E., Berroir, J. M., Delalande, C., Magnea, N., Mariette, H., Allegre, J., andCalatayud, J. *Physical Review B*, 45(1992)6305.

[41]Zheng, R., Matsuura, M., and Taguchi, T. *Physical Review B*, 61(2000) 9960.

[42]Boujdaria, K., Ridene, S., and Fishman, G. Luttinger-like parameter calculations. *Physical Review B*, 63(2001) 235302.

[43]C. Kittel, "Quantum theory of solids, Dunod". (1967).

[44]Bryant, G. W. *Physical Review Letters*, 59(1987) 1140.

[45]Bryant, G. W. *Physical Review B*, 37(1988) 8763.

[46]Saxena, A. K. *Solid state communications*, 113 (1999) 201-206.

[47]Pryor, C. *Physical Review B,* 57(1998) 7190.

[48]Brusaferri, L., Sanguinetti, S., Grilli, E., Guzzi, M., Bignazzi, A., Bogani, F.,andFranchi, S.*Applied physics letters,* 69(1996)3354-3356.

[49]Lobo, C., Perret, N., Morris, D., Zou, J., Cockayne, D. J. H., Johnston, M. B., and Leon, R. *Physical Review B*, 62(2000)2737.

[50]Xu, Z. Y., Lu, Z. D., Yang, X. P., Yuan, Z. L., Zheng, B. Z., Xu, J. Z., and Chang, L. L.. *Physical Review B*, 54(1996) 11528.

[51]El-Yadri, M., Aghoutane, N., Feddi, E., andDujardin, F. *Superlattices and Microstructures,* 102(2017)382-390.

[52]López, S. Y., Porras-Montenegro, N., & Duque, C. A. physica status solidi (b), 246(2009) 630-634.

CHAPTER 2: STUDYING THE EFFECTS OF TEMPERATURE AND PRESSURE ON EXCITONIC STATES IN A CYLINDRICAL QUANTUM BOX OF FINITE CONFINEMENT POTENTIAL

I-Introduction

The development of heterostructures in semiconductor materials has provided clear illustrations of certain quantum mechanical concepts, such as confinement and the quantization of energy levels. Advances in growth techniques have made it possible to realize different forms of semiconductor nanostructures (wells, wires and quantum boxes) so as to confine carriers in one, two or three directions in space.

BQs are of great importance by virtue of the fact that the motion of charge carriers is confined to three dimensions of space, leading us to new electronic properties and phenomena with vast potential applications in optoelectronics [1-4]. Theoretical and experimental work has also been devoted to the qualitative understanding of the properties of excitons and impurities in quantum wells, quantum rows and BQs [5-10].

In a semiconductor, an electron-hole pair can be created by absorption of a photon whose energy is greater than the band gap. This pair can reduce its energy by binding to form a stable state called an exciton, with a lifetime of between $10^{-9}\ s\ et\ 10^{-3}\ s$. The pair thus formed can move through the crystal, carrying excitation energy but no charge. There are two different limit approximations to the exciton. One is due to Frenkel (1931), who considers that the hole and electron are closely linked, and the exciton is said to be localized. The other is due to Mott and Wannier (1937), for whom the bond is weak and the distance between the electron and the hole is large in relation to the crystalline parameter; in this case, the exciton extends over several atomic sites. There is, however, an intermediate case, found in organic molecular crystals, where the distance between the electron and the hole corresponds to one or two times the intermolecular distance of the nearest neighbor; this exciton is called a charge-transfer exciton. Another way of describing an exciton is to see it as a wave of neutral polarization in the material.

Excitonic properties in quantum nanostructures have been intensively studied in recent times by virtue of their great importance in fundamental and applied physics. Among these nanostructures, QDs are three-dimensional wells that can trap electrons and holes resulting in quantized energy levels. The density of states is similar to a gamma function, and the particle wave function is localized inside the box. Excitons therefore play a key role in optical properties, and their stability is important for devices requiring this feature.

It is well established that the confinement of excitons in BQs produces an enhanced excitonic effect, which can be exploited in applications relating to new optoelectronic devices. Exciton binding energy increases with decreasing BQ dimensions, reaching a maximum around a critical width and then decreasing. This implies the existence of a critical confinement limit for which the effect of quantum confinement is large. There has been growing interest in the subject of confined exciton states in various nanostructures [11-19]. The reason behind this interest lies in the possible simplification of the heavy calculations involved in studying exciton binding energies in quantum structures such as QB systems. A high binding energy allows excitonic states to become stable. This stability enables excitons to be exploited in a wide range of applications, even at room temperature. With the aim of improving the physical properties of excitons and impurities in BQs, even at room temperature, several works have investigated the effect of pressure and temperature on excitons in nanostructures [20-24].

Raigoza et al [20] studied the effects of hydrostatic pressure on excitonic states in the$GaAs/Ga_{1-x}Al_xAs$ semiconductor, showing good agreement with experiment. Excitonic states in the$GaAs/Ga_{1-x}Al_xAs$ system under the influence of an electric field and hydrostatic pressure were also examined in this work. Oyoko et al [22] studied donor impurities in parallelepiped-shaped BQs of$GaAs$ - ( )GaAl As and found that the donor binding energy increases with pressure and decreases with BQ size. Kasapoglu [23] examined the combined effects of hydrostatic pressure and temperature on the binding energies of donor impurities in the$GaAs/Ga_{0.7}Al_{0.3}As$ based double quantum well. They showed that an increase in temperature gives rise to a decrease in the binding energy of the donor impurity, while a decrease in isothermal pressure improves the binding energy. More recently, Duque et al [24] investigated the combined effects of electron-hole correlation, hydrostatic pressure and temperature on third harmonic generation in parabolic BQs of $GaAs$ disk-shaped using the density matrix formalism and effective mass approximation.

In this chapter, we have examined the combined effect of temperature and pressure on the excitonic energy spectrum in a QB of cylindrical geometry. To obtain qualitative results as a solution to this problem, we adopted the variational approach in our calculation, while introducing the effective mass approximation to obtain the ground state energy as well as the binding energy of the$GaAs$ material embedded in another $Ga_{1-x}Al_xAs$ barrier material. We have chosen the excitonic confinement potential to be a finite potential. We are required to theoretically solve the Schrödinger equation of the system. In the following, we present the theoretical model and the results of a numerical calculation, accompanied by a detailed discussion.

II-Variational calculation of the binding energy of the exciton

1-Single-particle states

Let's consider an exciton confined in a cylindrical QB of radius R and height H, encapsulated by another barrier semiconductor of larger gap width. Within the framework of finite confinement potential approximations, we assume that the band offsets of the two materials are sufficiently small that the envelope function approximation can be used in a two-band model of the effective mass. The single-particle effective Hamiltonian has the expression :

$$H = -\frac{\hbar^2}{2m_i^*}\nabla_i^2 + V_\omega^i(r_i) \qquad pour \ (i = e, h), \tag{2.1}$$

where m_e^* et m_h^*, represent the effective masses of the electron and hole, respectively. V_ω^e and V_ω^h are the quantum confinement potential of the electron and hole, respectively, resulting from the band gap between the two semiconductors:

$$V_w^i = \begin{cases} 0 & \text{à l'intérieure de la boite} \\ v^i & \text{à l'extérieurede la boite} \end{cases} . \tag{2.2}$$

We choose z_e and z_h as the coordinates of the electron and hole along the z axis, chosen according to the axis of symmetry of the cylinder revolution, while ρ_e and ρ_h are the polar coordinates of the electron and hole in a plane perpendicular to the cylinder axis. With this parameterization, the confinement potential can be expressed analytically as :

$$V_w^i(\rho_i, z_i) = V_i\theta[\rho_i - R]\theta\left[|z_i| - \frac{H}{2}\right] \quad (i = e, h) \tag{2.3}$$

where V_ω^e and V_ω^h are respectively the heights of the potential barriers relative to the electron and   to the hole, and θ the Heaviside jump function.

In the case we are interested in here, where lateral and axial confinements coexist, the functions f_i and g_i construisent notre fonction d'essai, defined by :

$$f_i(\rho_i) = \begin{cases} J_0\left(\theta_i \frac{\rho_i}{R}\right) & pour & \rho_i \leq R \\ A_i K_0(\beta_i \rho_i) & pour & \rho_i \geq R \end{cases} \quad (i = e, h). \qquad (2.4)$$

$$g_i(z_i)_{i=} \begin{cases} \cos\left(\pi_i \frac{z_i}{H}\right) & si & |z_i| < \frac{H}{2} \\ B_i \exp(-K_i|z_i|) & si & z_i \geq \frac{H}{2} \end{cases} \quad (i = e, h). \qquad (2.5)$$

Let's rewrite the confinement potential as :

$$V_\omega^i(r_e) = V_{w\rho}^i(\rho_i) + V_{wz}^i(z_i) - \delta V_i(\rho_i, z_i), \qquad (2.6)$$

où $V_{w\rho}^i$ et V_{wz}^i are, respectively, the lateral and axial confinement potentials, while while δV_i is a corrective term; they are defined by :

$$V_{w\rho}^i(\rho_i) = -)V_i\theta(\rho_i R \quad (i = e, h). \qquad (2.7)$$

$$V_{wz}^i(z_i) = -]V_i\theta[|z_i|\frac{H}{2}(i = e, h). \qquad (2.8)$$

$$\delta V_i(\rho_i, z_i) = \begin{cases} 0 & si \ \rho_i \leq R \ et |Z_i| \leq \frac{H}{2} \\ V_i & sinon \end{cases}. \qquad (2.9)$$

If we consider $-\delta V_i$ as a perturbation potential, the energy of the particle in the ground state can be approximated by :

$$E_i = E_{i\rho} + E_{iz} - \langle \delta V_i \rangle$$

$$= \tau_i[(\theta_i/R)^2 + (\pi_i/H)^2] \frac{\langle f_i g_i | \delta V_i | f_i g_i \rangle}{\langle f_i g_i | f_i g_i \rangle} - . \qquad (2.10)$$

For large cylinders, the corrective term $d\hat{u}$ at $-\delta V_i$ cancels out. In the limit, expression (2.9) reduces to equation :

$$E_i = \tau_i[(\theta_i/R)^2 + (\pi_i/H)^2] \qquad (2.11)$$

$$\text{avec} \qquad \tau_i = \frac{\hbar^2}{2m_i^*} \qquad (i = e, h).$$

$$(2.12)$$

2-Effective system **Hamiltonian**

In ***Figure 2.1*** we designate our model composed of a BQ of cylindrical geometry based on a semiconductor material of type $GaAs/Ga_{1-x}Al_xAs$ with their confinement potentials corresponding to mole fractions of aluminum . x

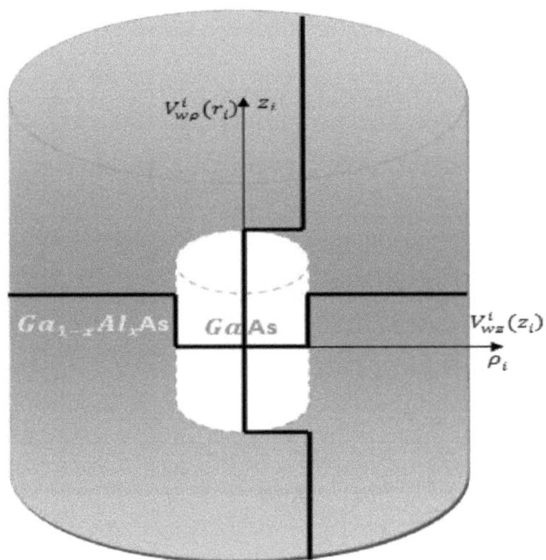

Figure 2.1 *Schematic diagram of a cylindrical BQ with axial and lateral confinement potential.*

We will neglect the electron-phonon coupling, as well as the electron-hole exchange interaction. In the case of non-degenerate parabolic bands, the Hamiltonian of such a system under external hydrostatic pressure P and temperature T, can be written as follows:

$$H_{ex}(r_e, r_h, P, T) = -\frac{\hbar^2}{2m_e^*}\nabla_e^2 - \frac{\hbar^2}{2m_h^*}\nabla_h^2 - \frac{e^2}{\varepsilon_0(P,T)|r_e - r_h|} + V_\omega^e(r_e, P, T) + V_\omega^h(r_h, P, T),$$

$$(2.13)$$

where m_e^* et m_h^*, represent the effective masses of the electron and hole respectively. And $\varepsilon_0(P,T)$ the dielectric constant of the can material under the effect of pressure and temperature. $r_e = (\rho_e, z_e)$ and $r_h = (\rho_h, z_h)$ represent the spatial coordinates of the electron and hole with respect to the cylinder's z axis, where z_e and z_h are the spatial coordinates of the electron and hole along the z axis, chosen according to the cylinder's axis of symmetry of revolution, while ρ_e and ρ_h are the respective distances of the electron and hole along the cylinder's axis.

$V_\omega^e(r_e, P, T)$ and $V_\omega^h(r_h, P, T)$ are the quantum confinement potential of the electron and hole respectively.

The two potentials $V_\omega^e(r_e, P, T)$ and $V_\omega^h(r_h, P, T)$ are written :

$$V_\omega^i(r_i, P, T) = \begin{cases} 0 & si \ \rho_i \leq R \ et \ |Z_i| \leq d \\ V_i(p, t) & sinon \end{cases} \quad \text{(i=e,h)} .$$

$$(2.14)$$

The expression $V_i(x, P, T)$ is obtained from the pressure-temperature dependence of the discontinuities in the two band gaps that make up the well and barrier materials. [25]

$$V_e(x, P, T) = 0.658 \Delta E_g(x, P, T), \qquad (2,15)$$

$$V_h(x, P, T) = 0.342 \Delta E_g(x, P, T), \qquad (2.16)$$

where

$$\Delta E_g(x, P, T) = \Delta E_g(x) - P(1.3 * 10^{-3})x - T(1.11 * 10^{-4})x, \qquad (2.17)$$

with

$$\Delta E_g(x) = E_{g\,Ga(1-x)Al(x)As}(x) - E_{g\,GaAs}(x) = 1.155x + 0.37x^2. \qquad (2.18)$$

The Coulombian interaction is written $\frac{\varepsilon_0(0)}{\varepsilon_0(P,T)} \frac{e^2}{|r_e - r_h|}$, where $\varepsilon_0(0)$ and $\varepsilon_0(P, T)$ are, respectively, the dielectric constant without and with pressure and temperature, and is given by [25-27] :

$$\varepsilon_0(P, T) =$$

$$\begin{cases} 12.7 \exp(-1.67 \times 10^{-3}P) \exp\big(9.4 \times 10^{-5}(T - 75.6)\big) & for \ 0 \leq T \leq 200 \ K \\ 13.8 \exp(-1.73 \times 10^{-3}P) \exp\big(20.4 \times 10^{-5}(T - 300)\big) & for \ T > 200 \ K \end{cases} .$$

$$(2.19)$$

Since temperature and pressure are external stresses, the gap width of the material $GaAs$ depends on these two stresses at the center Γ of the Brillouin zone. It can be expressed by the following relationship [26- 28]:

$$E_g(P, T) = 1.519 - \frac{5.4 \times 10^{-4}T^2}{T + 204} + 0.01261 \ P + 3.77 \times 10^{-5}P^2.$$

$$(2.20)$$

It is well known that hydrostatic pressure has a considerable effect on the dimensions of solid materials. *Yu et Cardona* [29] have shown that, in low-dimensional semiconductors, the size of BQs is also affected by deformation effects. They showed that the box volumes with and without pressure are linked by elastic constants. This description can be generalized to any nanostructure shape, assuming that the box radius follows the relationship:

$$R(P) = R(0)[1 - (S_{11} - 2S_{12})P]^{1/3}, \qquad (2.21)$$

R (0) is the pressure-free radius .$(P = 0)$$S_{11}$ and S_{12} are the compliance parameters linked to the elastic constants C_{11} andC_{12} by the relations below [30-32] :

$$S_{11} = \frac{c_{11}+c_{12}}{(c_{11}-c_{12})(c_{11}+c_{12})} \text{ and } . S_{12} = \frac{c_{12}}{(c_{11}-c_{12})(c_{11}+c_{12})} \qquad (2.22)$$

The effective Hamiltonian can be simplified using the system of atomic units. We take as unit of length the effective Bohr radius $a_{ex}^* = \frac{\varepsilon_0 \hbar^2}{\mu e^2}$, as unit of energy the effective

Rydberg$R_{ex}^* = \frac{\mu e^4}{2\varepsilon_0^2 \hbar^2}$, the effective Hamiltonian is thus written :

$$H_{eff} = -\frac{1}{1+\sigma}\frac{m_e^*}{m_e^*(P,T)}\left[\frac{\partial^2}{\partial \rho_e^2} + \frac{1}{\rho_e}\frac{\partial}{\partial \rho_e} + \frac{\rho_{eh}^2+\rho_e^2-\rho_h^2}{\rho_e\rho_{eh}}\frac{\partial^2}{\partial \rho_e \partial \rho_{eh}} + \frac{\partial^2}{\partial z_e^2}\right] - \frac{\sigma}{1+\sigma}\frac{m_h^*}{m_h^*(P,T)}\left[\frac{\partial^2}{\partial \rho_h^2} + \right.$$

$$\frac{1}{\rho_h}\frac{\partial}{\partial \rho_h} + \frac{\rho_{eh}^2+\rho_h^2-\rho_e^2}{\rho_h\rho_{eh}}\frac{\partial^2}{\partial \rho_h \partial \rho_{eh}} + \left.\frac{\partial^2}{\partial z_h^2}\right] - \left[\frac{\partial^2}{\partial \rho_{eh}^2} + \frac{1}{\rho_{eh}}\frac{\partial}{\partial \rho_{eh}}\right] - \frac{\varepsilon_0(0,300)}{\varepsilon_0(P,T)}\frac{2}{\sqrt{\rho_{eh}^2+(z_e-z_h)^2}} +$$

$$V_\omega^e(\rho_e, z_e, P, T) + V_\omega^h(\rho_h, z_h, P, T),$$

$$(2.23)$$

where $\sigma = \frac{m_e^*}{m_h^*}$ is the ratio of the effective masses of the electron and the hole.

3-Ecitonic wave function

In general, we can express the excitonic envelope function using the six independent coordinates ($\rho_i, \phi_i z_i$), whereϕ_i is the angular component of particle i (electron or hole). In the ground state, the system is rotationally invariant around the z-axis, so the angular dependence can be a function of the distance of the projections between electron and hole positions in the plane.

The pressure- and temperature-dependent excitonic wave function is written as [33] :

$$\psi_{ex}(\rho_e, \rho_h, z_e, P, T) = F_e(\rho_e, z_e, P, T)F_h(\rho_h, z_h, P, T)F_{eh}(\rho_{eh}, |z_e - z_h|), \qquad (2.24)$$

with :

$$F_{eh}(\rho_{eh}, |z_e - z_h|) = exp(-\alpha\rho_{eh}) \, exp\left[-\gamma(z_e - z_h)^{\frac{1}{2}}\right],$$

$$(2.25)$$

and

$$F_i(\rho_i, z_i, P, T) = f_i(\rho_i, P, T)g_i(, z_i, P, T), \qquad (2.26)$$

where $f_i(\rho_i, P, T)$ and $g_i(, z_i, P, T)$ are obtained by solving the two Schrödinger equations corresponding to the following two lateral and axial dimensions:

$$\begin{cases} \left[-\dfrac{\hbar^2}{2m_i^*(P,T)}\nabla_i^2 + V_\omega^i(\rho_i, P, T)\right] f_i(\rho_i, P, T) = E_i(\rho_i, P, T)f_i(\rho_i, P, T) \\ \left[-\dfrac{\hbar^2}{2m_i^*(P,T)}\nabla_i^2 + V_\omega^i(z_i, P, T)\right] g_i(z_i, P, T) = E_i(z_i, P, T)g_i(\rho_i, P, T) \end{cases}.$$

$$(2.27)$$

The solutions of this system of differential equations are of the following form :

$$f_i(\rho_i, P, T) = \begin{cases} J_0\left(\theta_i \dfrac{\rho_i}{R}\right) & pour & \rho_i \leq R \\ A_i K_0(\beta_i\rho_i) & pour & \rho_i \geq R \end{cases} (i = e, h),$$

$$(2.28)$$

$$g_i(z_i, P, T) = \begin{cases} cos\left(\pi_i \dfrac{z_i}{H}\right) & si & |z_i| < \dfrac{H}{2} \\ B_i exp(-K_i|z_i|) & si & z_i \geq \dfrac{H}{2} \end{cases} (i = e, h)$$

$$(2.29)$$

where j_0 and K_0 are modified Bessel functions of order 0. The parameters , $\theta_i(T, P)\pi_i(T, P)$, $A_i(T, P)$ and $B_i(T, P)$ are determined from the boundary conditions à $\rho_i = R$ and $|z_i| = \dfrac{H}{2}$. In equation (2.13), α and γ are two variational parameters introduced to account for the anisotropy of the box.

$K_{1i} = \sqrt{\dfrac{V_i}{\tau_i} - \left(\dfrac{\pi_i}{H}\right)^2}, \beta_i = \sqrt{\dfrac{V_i}{\tau_i} - \left(\dfrac{\theta_i}{R}\right)^2}$ are the axial and radial wave vectors respectively.

The continuity of the excitonic wave function and its first derivative at the interface $\rho_i = R$, allows us to write :

$$A_i = \frac{J_0(\theta_i)}{K_0(\beta_iR)}\frac{\theta_iJ_1(\theta_i)}{J_0(\theta_i)} = \frac{\beta_iRK_i(\beta_iR)}{K_0(\beta_iR)}. \qquad (2.30)$$

Similarly for the constants π_i and B_i , they are determined by the two previous conditions at the interface $z_i = \dfrac{H}{2}$:

$$B_i = cos\left(\frac{\pi_i}{2}\right) / exp\left(-\frac{K_iH}{2}\right). \qquad (2.31)$$

$$tan\left(\frac{\pi_i}{2}\right) = K_i\left(\frac{H}{\pi_i}\right) . \qquad (2.32)$$

In our work, we are interested in the symmetrical confinement potential, which is expressed as follows:

$$V_w^i(r_i, P, T) = V_{wp}^i(\rho_i, P, T) + V_{wz}^i(z_i, P, T) - \delta V_i(\rho_i, z_i, P, T), \qquad (2.33)$$

where V_{wp}^i and V_{wz}^i are the lateral and axial confinement potentials respectively, while δV_i is a corrective term. They are defined by :

$$V_{wz}^i(\rho_i, P, T) = V_i\,\theta(\rho_i - R) \; ; \quad V_{\omega z}^i(z_i, P, T) = V_i\theta\left(|z_i| - \frac{H}{2}\right), \qquad (2.34)$$

where θ is the Heaviside function.

$$\delta V_i = \begin{cases} 0 & si \ \theta_i < R \ ou \ |z_i| < H/2 \\ V_i & ailleurs \end{cases} . (2.35)$$

4-Exciton binding energy

The energy of the ground-state exciton E_f is determined on the basis of the variational principle by minimizing the average energy with respect to the variational parameters α and γ, i.e. :

$$E_f(\text{P, T}) = \min_{\alpha,\gamma}\langle\psi_{ex}|H_{eff}|\psi_{ex}\rangle. \qquad (2.36)$$

$$\langle E(\alpha,\gamma)\rangle = \frac{\langle\psi_{ex}|\hat{H}|\psi_{ex}\rangle}{\langle\psi_{ex}|\psi_{ex}\rangle} = \frac{1}{1+\sigma}\left[\frac{m_e^*}{m_e^*(P,T)}\left(\frac{\theta_e}{R}\right)^2 + \frac{m_h^*}{m_h^*(P,T)}\sigma\left(\frac{\theta_h}{R}\right)^2\right] - \alpha^2 + \alpha\frac{P_3(\alpha,R)}{P_1(\alpha,R)} -$$

$$\frac{\alpha}{1+\sigma}\frac{\frac{m_e^*}{m_e^*(P,T)}P_4(\alpha,R) + \frac{m_h^*}{m_h^*(P,T)}\sigma P_5(\alpha,R)}{P_1(\alpha,R)} + 2\gamma - 4\gamma^2\frac{Z_2(\gamma,H)}{Z_1(\gamma,H)} + \frac{4\gamma^2}{1+\sigma}\frac{\frac{m_e^*}{m_e^*(P,T)}Z_3(\gamma,H) + \frac{m_h^*}{m_h^*(P,T)}\sigma Z_4(\gamma,H)}{Z_1(\gamma,H)} +$$

$$\langle V_{coul}\rangle - V_e\left(1 - \frac{P_7(\alpha,R)}{P_1(\alpha,R)}\right)\left(1 - \frac{Z_5(\gamma,H)}{Z_1(\gamma,H)}\right) - V_h\left(1 - \frac{P_8(\alpha,R)}{P_1(\alpha,R)}\right)\left(1 - \frac{Z_6(\gamma,H)}{Z_1(\gamma,H)}\right).$$

(2.37)

The subband energy $E_l(P,T)$ is defined as the sum of the electron and hole energies without Coulomb interaction:

$$E_l = E_e + E_h \qquad (2.38)$$

It is obtained by solving the Schrödinger equation, while neglecting the electron-hole correlation, i.e. the variational parameters α et γ sont nul and the coulombic interaction term.

The integrals P_i are defined by :

$$P_i = 8\pi \int_0^\infty d\rho_e \int_0^\infty d\rho_h \int_{|\rho_e - \rho_h|}^{\rho_e + \rho_h} \frac{F_i(\rho_e, \rho_h, \rho_{eh})\rho_e\rho_h\rho_{eh}}{\sqrt{[(\rho_e + \rho_h)^2 - \rho_{eh}^2][\rho_{eh}^2 - (\rho_e + \rho_h)^2]}} d\rho_{eh},$$

$$(2.39)$$

$$F_1(\rho_e, \rho_h, \rho_{eh}) = [F_e(\rho_e)F_h(\rho_h)\exp(-\alpha\rho_{eh})]^2, \qquad (2.40)$$

$$F_3(\rho_e, \rho_h, \rho_{eh}) = \frac{1}{\rho_{eh}} F_1(\rho_e, \rho_h, \rho_{eh}) \tag{2.41}$$

$$F_4(\rho_e, \rho_h, \rho_{eh}) = -\frac{\rho_{eh}^2 + \rho_e^2 - \rho_h^2}{\rho_e \rho_{eh}} f_e(\rho_e) f_e'(\rho_h)[f_h(\rho_h)\exp(-\alpha\rho_{eh})], \tag{2.42}$$

$$F_5(\rho_e, \rho_h, \rho_{eh}) = -\frac{\rho_{eh}^2 + \rho_h^2 - \rho_e^2}{\rho_h \rho_{eh}} f_h(\rho_e) f_h'(\rho_h)[f_e(\rho_h)\exp(-\alpha\rho_{eh})], \tag{2.43}$$

$$F_7(\rho_e, \rho_h, \rho_{eh}) = F_1(\rho_e, \rho_h, \rho_{eh})\theta(R - \rho_e), \tag{2.44}$$

$$F_8(\rho_e, \rho_h, \rho_{eh}) = F_1(\rho_e, \rho_h, \rho_{eh})\theta(R - \rho_h). \tag{2.45}$$

The Pi integrals are evaluated numerically after performing the transformations and variable changes shown in Appendix A.

The Zi integrals are given by :

$$Z_i = \int_{-\infty}^{\infty} dZ_e \int_{-\infty}^{\infty} dZ_h \, G_i(Z_e Z_h), \tag{2.46}$$

$$G_1(z_e, z_e) = [g_e(z_e)g_h(z_h)exp(-\gamma(z_e - z_h)^2)] \tag{2.47}$$

$$G_2(z_e, z_e) = (z_e - z_h)^2 G_1(z_e, z_e), \tag{2.48}$$

$$G_3(z_e, z_e) = (z_e - z_h)g_e(z_e)g_e'(z_e)[g_h(z_h)exp(-\gamma(z_e - z_h)^2)], \tag{2.49}$$

$$G_4(z_e, z_e) = (z_e - z_h)g_h(z_h)g_h'(z_h)[g_e(z_e)exp(-\gamma(z_e - z_h)^2)], \tag{2.50}$$

$$G_5(z_e, z_e) = G_1(z_e, z_e)\theta(d - |z_e|), \tag{2.51}$$

$$G_6(z_e, z_e) = G_1(z_e, z_e)\theta(d - |z_h|). \tag{2.52}$$

For the calculation of Coulomb energy, we have decided to consider the approximation performed by Le Goff [34] to avoid any complication in the numerical evaluation of quintuple integrals.

$$\langle V_{coul} \rangle = -2 \int_{-\infty}^{\infty} dz_e \int_{-\infty}^{\infty} G_1(z_e, z_h)P_c(\alpha, R, z_e - z_h)dz_h. \tag{2.53}$$

P_c is an integral of type P_i with a function F_c to be integrated, defined by :

$$F_c(\rho_e, \rho_h, \rho_{eh}) = \frac{F_1(\rho_e, \rho_h, \rho_{eh})}{\sqrt{\rho_{eh}^2 + (z_e - z_h)^2}}. \tag{2.54}$$

We determine the Coulomb energy using the following approximate expression:

$$\Phi(z_e - z_h) = \frac{A}{B + |z_e - z_h|} \tag{2.55}$$

where the constants A and B are chosen such that the correct values of $< V_{coul} >$ are reached within the limits $|z_e - z_h| \rightarrow 0$ and $|z_e - z_h| \rightarrow \infty$.

If$| ze - zh |$ is large,$\Phi(ze - zh)$ is given by :

$$\Phi\left(z_e - z_h\right) = \frac{A}{|z_e - z_h|}.$$

$$(2.56)$$

Pc then becomes :

$$P_c(\alpha, R, z_e - z_h) = \frac{P_1(\alpha, R)}{|z_e - z_h|}.$$

$$(2.57)$$

We can therefore deduce the value of the constant A, *which is written as* :

$$A = P_1(\alpha, R).$$

$$(2.58)$$

On the other hand, for small$|z_e - z_h|$ values:

$$\Phi\left(z_e - z_h\right) = \frac{A}{B}.$$

$$(2.59)$$

WhileP_c is written

$$P_c(\alpha, R, z_e - z_h) = P_3(\alpha, R)$$

$$(2.60)$$

which gives :

$$\frac{A}{B} = P_3(\alpha, R).$$

$$(2.61)$$

Hence the approximate expression of the Colombian potential is written as follows:

$$< V_{coul} >= -2\frac{Z_c(\alpha, R, \gamma, d)}{Z_c(\gamma, H)}$$

$$(2.62)$$

with the Zc integral is analogous to the Zi integrals where :

$$G_c(R, d, \alpha, \gamma, z_e, z_h) = \frac{G_1(z_e, z_h)}{\frac{P_1(\alpha, R)}{P_3(\alpha, R)} + |z_e - z_h|}.$$

$$(2.63)$$

The dependence of the excitonic ground-state binding energy on temperature and pressure is defined as the difference between the sum of the free-state electron and hole energies and the exciton ground-state energy. It can be written as :

$$E_B(P, T) = E_l(P, T) - E_f(P, T).$$

$$(2.64)$$

5-Calculation of photoluminescence energy

Photoluminescence is a powerful optical technique for characterizing semiconductor materials and insulators. Its operating principle is simple, as shown in

Figure 2.2. The electrons of the substance under study are excited by radiation (usually monochromatic), and the light emitted by the substance is detected. In general, the energy of the light emitted is lower than that of the radiation used for excitation. In practice, the intensity emitted by solids is often very low. It is therefore necessary to use a laser as an excitation , together with a high-performance detection system.

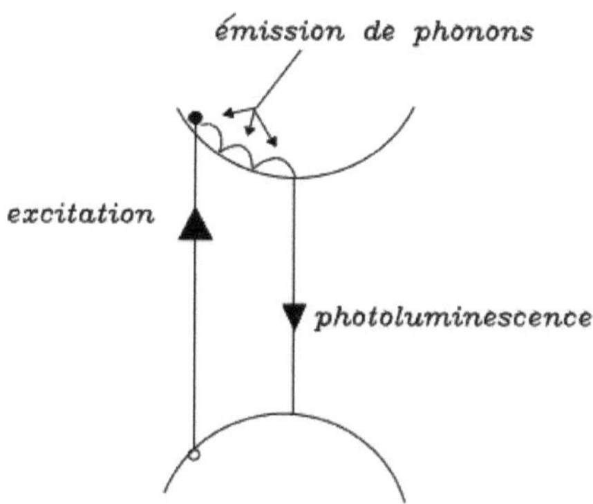

Figure 2.2: shows the principle of photoluminescence energy emission.

The photoluminescence energy is defined as the sum of the binding energy and the gap energy of the material:

$$E_{pl}(P,T) = E_B(P,T) + E_{GAP}(P,T). \tag{2.65}$$

III-Numerical resolution method

The variational method used in this work consists in considering the state of the system as having a minimum energy, to which correspond very precise values of variational parameters α et γ. To determine α_{min} et γ_{min} we fix the dimensions of the BQ, then use the iterative method to determine the corresponding parameters and then obtain the minimum energy. The various energy terms (integrals, special functions, etc.) are calculated numerically. The flow chart in Appendix C summarizes the various steps involved in numerically determining the binding energy.

IV-Numerical results and discussion

The application of external stresses (pressure and temperature) modifies the latent conditions, barrier height, box dimensions and dielectric constant. The barrier potential can be obtained using equations (2.14), (2.15) and (2.16). The variation of the dielectric constant as a function of pressure is given by equation (2.18) with P in GPa . The variation of the box dimension with pressure is taken into account using equation (2.19).

For our numerical results, we have chosen as an application example a cylindrical BQ based on $GaAs$ encapsulated by a barrier semiconductor of type $GaAs/Ga_{1-x}Al_x$. The corresponding physical parameters are shown in the table below:

$m_e^* = 0.063\ m_0$	$m_h^* = 0.079\ m_0$	$\varepsilon_0(0,300) = 13.18$
$E_g^\Gamma(0,300)$ $= 1.422eV$	S_{11} $= 1.16 \times 10^{-2} GPa^{-1}$	$a_{ex}^* = 19.6\ nm$
$\Delta_0 = 0.341eV$	$S_{12} = -3.7 \times 10^{-3} GPa^{-1}$	$R_{ex}^* = 2.78\ meV$

Table 2.1: *The $GaAs$ [35] parameters used in our calculations.*

Figure 2.3 shows the electron wave function for various values of electron position in the nucleus. This figure shows that the electron wave function is mainly confined to the central region and, although we have used a finite confinement potential, the presence of electrons in the shell region is practically marginalized. This is caused by the strong confinement due to the large conduction band offset between the core and barrier materials. For small values of radius R, discrete electron levels are absent in the well, and the electron and hole wave functions are distributed in the barrier materials. By increasing the radius R of the cylinder, the electron energy levels fall onto the continuous spectrum in the well.

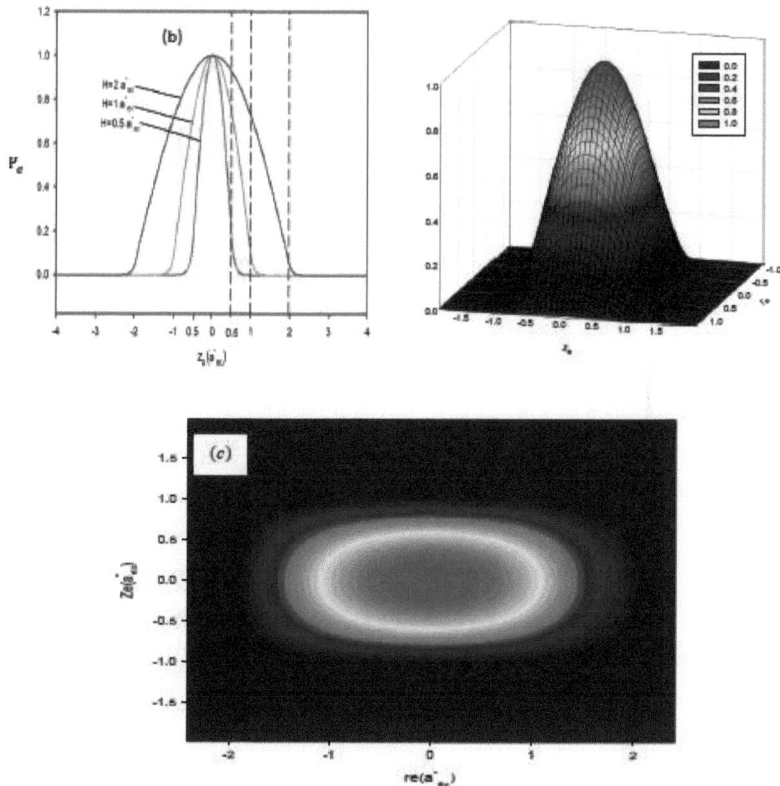

Figure 2.3: *Variation of the electron wave function as a function of its position in the quantum box.*

Figure 2.4a shows the exciton binding energy as a function of pressure for different temperatures and aluminum concentrations in the barrier material. It is clear that the binding energy increases linearly with pressure, whatever the concentration x and temperature. This means that the exciton remains stable even at room temperature. This behavior justifies the fact that the effects of stress induce additive confinement in addition to geometric confinement. On the contrary, the opposite behavior can be seen in ***Figure 2.4b***, where we plot the variation of binding energy as a function of T. These curves show that the binding energy decreases almost linearly as a function of T, following two negative slopes, whatever the value of pressure and can size.

In order to give a simple quantitative description of the variation of binding energy as a function of pressure and temperature, we have established a simple analytical form of excitonic binding energy in terms of P and T. Since the variation increases with a positive slope of the binding energy E_b as a function of P, we can write the law of variation of E_b as a function of pressure P under the equation $E_b(P) = E_b(0) + 0.011P$. The same reasoning is applied to the case of temperature, where the variation in binding energy follows two analytical equations:

$$\begin{cases} E_b(T) = E_b(0) - 0.000422T \ pour\ 0 < T < 200 \\ E_b(T) = E_b(200) - 0.00087T \ pour \quad T > 200 \end{cases}$$

In *Figure 2.5*, we show the variation in exciton photoluminescence energy as a function of can radius R and height H for different pressure values (0 and 40 kbar) at a fixed temperature and mole fraction of aluminum, respectively: $T = 300K \ et \ x = 0.3$. This illustration has enabled us to clearly envisage the conditions under which photoluminescence energy is maximal. An initial analysis of this curve shows that, for a given pressure, photoluminescence energy is sensitive to variations in can size (radius and height).

A rapid decrease in this emission is observed as the size increases, and retains the same behavior for both values of pressure, meaning that varying pressure significantly affects photoluminescence emission, and it becomes more important for smaller box sizes. The practical interest of these effects is that they can be exploited to modify the optical properties linked to the existence of excitons at a given temperature without changing the size of the BQs but simply by controlling the external pressure. It is important to note that this figure also shows that for a fixed value of R and H, the energy of the exciton decreases with increasing T, which tends to reduce the electron-hole attraction, and consequently the exciton loses its stability.

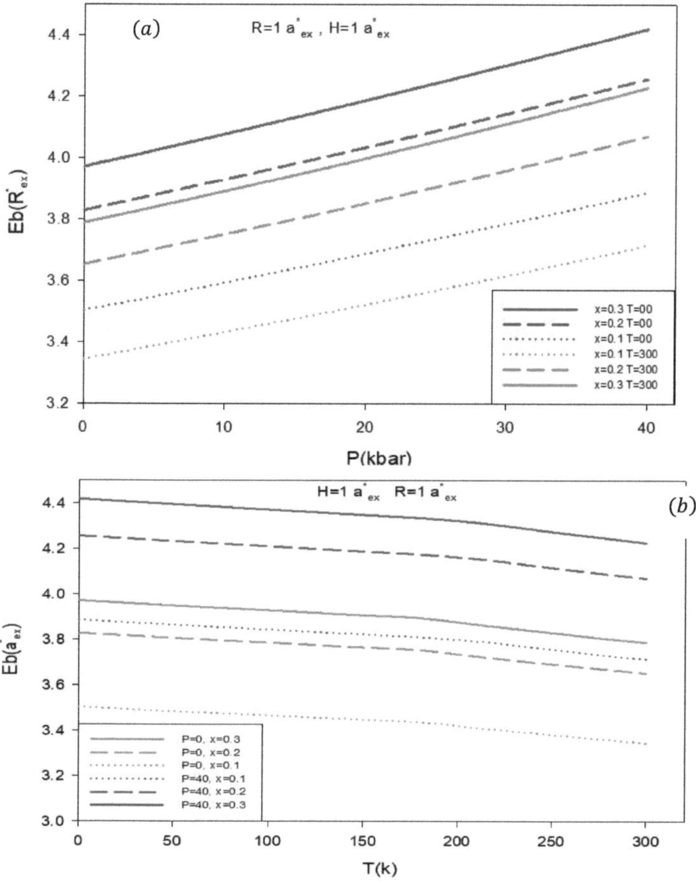

Figure 2.4- *Variation in binding energy for different values of aluminum concentration* $x(x = 0.1, 0.2, 0.3)$ *as a function of (a) pressure for different values of pressure, (b) temperature for different values of pressure.*

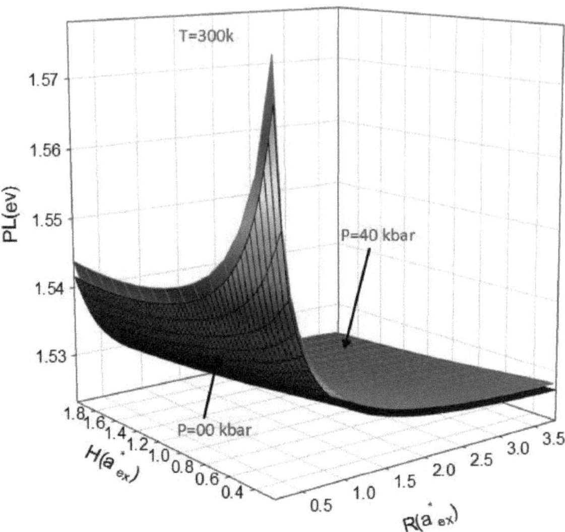

Figure 2.5: *Variation of exciton photoluminescence energy as a function of box radius R and height H for different pressure values (0kbar,40kbar) at a fixed temperature and aluminum mole fraction, respectively: .T = 300K, x = 0.3*

The above descriptions of the effects of pressure and temperature can be supported by analyzing the photoluminescence energy transition in *Figure 2.6*, where we have presented the variation defined by the PL equation as a function of temperature for different sets (P, R)= (0.0.8), (20.0.8), (40.0.8), (0.1), (20.1), (40.1), (0.1.5), (20.1.5) and (40.1.5). For each pressure, the photoluminescence energy follows a decreasing function with temperature and the radius R of the can. This means that applying temperature decreases the energy of the light emitted. On the other hand, for a fixed value of temperature T, an increase in applied pressure leads to an increase in emitted intensity, whatever the chosen radius

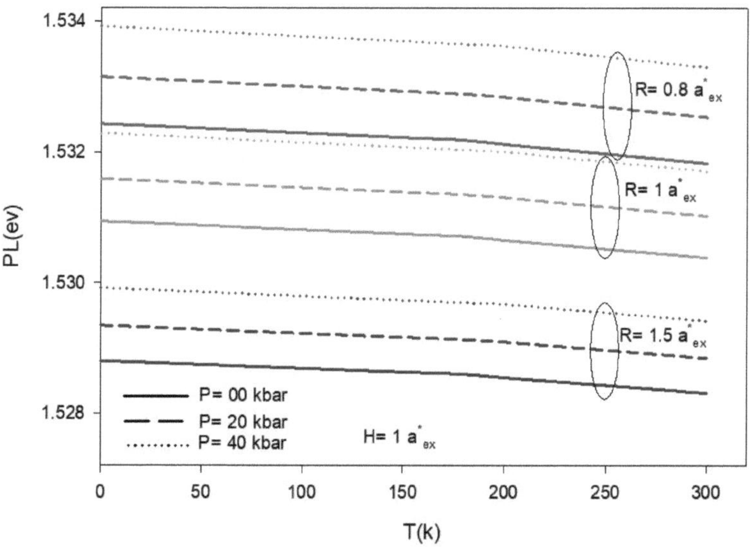

Figure 2.6: *Variation of photoluminescence energy as a function of temperature for different values of cylinder radius R and pressure at height H=1 .a_{ex}^{*}*

Conclusion

To conclude this chapter, we have established a detailed theoretical calculation based on the variational Ritz method to solve the Schrödinger equation of an exciton written in the effective mass approximation, which remains a more compact approximation and yields fairly convincing results for various problems. We have chosen the case of the finite potential barrier, to examine the behavior of the exciton's ground-state binding energy. With this choice of potential, the binding energy of the exciton is affected by the variation in the dimension of the QB (reducing the dimensionality of the system). Sensitivity to spatial confinement is greater for lateral confinement than for axial confinement. Varying the aluminum concentration of the barrier material has a significant influence on exciton binding energy, since it increases the height of the potential barrier, and has a remarkable effect on small BQs. We also investigated the combined contributions of temperature and pressure to exciton binding energy, on the $GaAs/$ $Ga_{1-x}Al_xAs$ heterostructure, characterized by an intermediate coupling constant. The results obtained show that the combined effects of these two external constraints play a fundamental role through its impact on the binding energy, which is remarkable for the different boundary structures, namely the disk, the quantum well and the quantum wire. We also point out that the effective mass approximation underlying our work is inadequate to describe the behavior of the excitonic binding energy for thin BQs, due to the effects of band non-parabolicity. In relation to optical properties, we have also shown that geometric confinement, aluminum concentration and two external constraints (pressure and temperature) have a significant effect on photoluminescence energy emission. Finally, we believe that through the above results, we have opened the way to the possibility of a good understanding and explanation of the various physical processes provided by the experimental results (optical properties, PL spectrum, absorption...), concerning the existence and stability of excitons in semiconductor-based BQs.

Bibliography

[1]D.H. Davies, "The physics of Low-Dimensional Semiconductors", Cambridge University Press, (1998).

[2]Harrison, P., and Valavanis, A. *"Quantum wells, wires and dots: theoretical and computationalphysics of semiconductor nanostructures"*. John Wiley& Sons (2016).

[3]C. Weisbuch and B. Vinter, "Quantum Semiconductor Structures: Fundamentals andApplications" (Academic, Boston, 1991).

[4]Klingshirn, Claus F. *Semiconductoroptics*. Springer Science and Business Media, (2012).

[5]U. C. Mendes, M. Korkusinski, A. H. Trojnar, and P. Hawrylak, *Phys. Rev. B* 88 (2013)115306.

[6]C. Gonzalez-Santander, F. Dominguez-Adame, *Physics Letters A* 374 (2010) 2259-2261.

[7]A. El Moussaouy, D. Bria, A. Nougaoui, R. Charrour and M. Bouhassoune, *J. Appl. Phys93*,(2003) 2906.

[8]E. Kasapoglu, C.A. Duque, S. Sakiroglu, H. Sari, and I. Sokmen, *Physica E* 43 (2011)1427-1432.

[9]A. Musial, P. Kaczmarkiewicz, G. Sek, P. Podemski, P.Machnikowski, J. Misiewicz, S. Hein, S. H?fling, and A. Forchel,*Phys. Rev. B* 85,(2012) 035314.

[10]M. Holmes, S. Kako, K. Choi, P. Podemski, M. Arita, and Y. Arakawa, *Phys. Rev. Lett.* 111,(2013) 057401.

[11]Y. Kayanuma, *Phys. Rev. B* 41 (1990) 10261.

[12]G. Einevoll, *Phys. Rev. B.* 45 (1992) 3410.

[13] J.L. Marin, R. Riera, and S.A. Cruz, J. *Phys. Condens. Matter* 10 (1998) 1349.

[14]C. Schulhauser, D. Haft, R. J. Warburton, K. Karrai, A. O. Govorov, A. V. Kalameitsev, A. Chaplik, W. Schoenfeld, J.M. Garcia, and P. M. Petroff, *Phys. Rev. B* 66, (2002) 193303.

[15]E.W.S. Caetano, V.N. Freire, G.A. Farias, and E.F. da Silva, *Brazilian J. Phys.* 34 (2004) 702.

[16]H. Hassanabadi, A.A. Rajabi, S. Zarrinkamar, and M.M. Sarbazi, *Few-Body Syst.* 45 (2009) 71.

[17]M. Santhi, and A.J. Peter, *Eur. Phys. J. B* 71 (2009) 225.

[18]P. Vasa, R. Pomraenke, S. Schwieger, Yu. I. Mazur, Vas. Kunets, P. Srinivasan, E. Johnson, J. E. Kihm, D. S. Kim, E.Runge, G. Salamo, and C. Lienau1, *Phys. Rev. Lett.* 101 (2008)116801.

[19]S.D. Wu and L. Wan, *Eur. Phys. J. B* 85 (2012) 12.

[20]N. Raigoza, C.A. Duque, N. Porras-Montenegro, and L.E. Oliveira, *Physica B* 371 (2006) 153.

[21]C. Duque, N. Poras- Montenegro, Z. Barticevic, M. Pacheco, and L. E. Oliveira, *J. Phys.: Condens. Matter* 18 (2006) 1877.

[22]H.O. Oyoko, C.A. Duque, and N. Porras-Montenegro, *J. Appl. Phys.* 90 (2001) 819.

[23]E. Kasapoglu, *Phys. Lett. A* 373 (2008) 140.

[24]C.M. Duque, M.E. Mora-Ramos and C.A. Duque.*J Nanopart Res* 13 (2011) 6103.

[25]R.F. Kopf, M.H. Herman, M.L. Schnos, A.P. Perley, G. Livescu, and M. Ohring, *J. Appl. Phys.* 71, 5004 (1992) 5004-5011.

[26]M. El-Yadri, N. Aghoutane, E. Feddi, and F. Dujardin, *Superlattices and Microstructures*, 102,(2017) 382-390.

[27]Z. Zeng, G. Gorgolis, C. S. Garoufalis, and S. Baskoutas, *Science advanced materials* 6 (2014)586-591.

[28]S. Y. Lopez, and N. Porras-Montenegro, C. A. Duque, *physicastatussolidi* (b) 246 (2009) 630-634.

[29]C. A. Duque, S. Y. López, and M. E. Mora-Ramos, *physicastatussolidi* (b) 244 (2007) 1964-1970.

[30]Y. Yu, and M. Cardona, "Fundamentals of Semiconductors", Springer, Berlin, (1998).

[31]F.J. Culchac, and N. Porras-Montenegro, A. Latge, *J. Appl. Phys.* 105 (2009) 094324.

[31]F.J. Culchac, N. Porras-Montenegro, A. Latge, J. Appl. Phys. 105 (2009) 094324.

[32]H. M. Baghramyan, M. G. Barseghyan, A. A. Kirakosyan, R. L. Restrepo, C.A. Duque, Journal of Luminescence 134 (2013) 594-599.

[33]A. El Moussaouy, N. Ouchani, Y. El Hassouani, and D. Abouelaoualim, *Superlattices and Microstructures* 73 (2014) 22-37.

[34]S. Le Goff and B. Stébé, *Phys. Rev.* **B47** (1993)1383.

[35]H. O. Oyoko, C. A. Duque, and N. Porras-Montenegro, *J. Appl. Phys.* 90 (2001) 819-823.

CHAPTER 3: STUDY OF EXCITONIC TRION PROPERTIES UNDER THE EFFECT OF TEMPERATURE IN A CYLINDRICAL QUANTUM BOX OF FINITE AND INFINITE CONFINEMENT POTENTIAL

I-Introduction

In this introduction we review the main experimental and theoretical work on excitonic trions in bulk semiconductors, quantum wells and quantum dots.

By analogy with certain small stable molecular and atomic systems, e.g. hydrogenoid edifices, , H^- H_2^+ and H_2 . Lampert [1] suggested the existence of several excitonic complexes in bulk semiconductors, simply by generalizing Wannier's exciton concept [2]. Excitonic complexes that can result from the binding of an exciton with neutral or ionized impurities are known as "localized complexes" or of an exciton with neutral or charged quasiparticles are "mobile complexes".

Among localized complexes, the binding of an exciton to an impurity leads, depending on the charge state of the impurity, to various neutral exciton-donor complexes $(D^0.X)$ neutral exciton-acceptor (A^0, X) , or ionized exciton-donor$(D^+.X)$, which were first detected in the photoluminescence spectra of Silicon [3] and other semiconductors [4-6].

The exciton is free to move in the crystal, but can also bind to another exciton to generate a neutral, mobile excitonic complex. The biexciton or excitonic molecule X_2 has been identified in $ZnSe$ [7], Germanium [8],$CuCl$ [9,10], and in quantum wells$GaAs/AlGaAs$ [11].

Charged excitons or excitonic trions are three-particle complexes that are mobile and charged and can be categorized into two types:

> ➤ positive trions$X_2^+(e, h, h)$ resulting from the Coulombic interaction between two holes and an electron.

> ➤ negative trions $X^-(e, e, h)$ resulting from the Coulomb interaction between two electrons and a hole.

Our study focuses on negative trions in a cylindrical BQ geometry based on $GaAs/AlGaAs$. Theoretical studies have shown the stability of excitonic trions in bulk semiconductors [12,13], and in two-dimensional media by variational calculation [14] or by analytical calculation [15] of the trion binding energy. Excitonic trions were observed by luminescence experiments for the first time in low-temperature bulk Germanium [16] and also by luminescence and absorption experiments in bulk$CuCl$ thin films [17]. Observations of trions in bulk semiconductors are generally unconvincing due to their low binding energy. Variational calculation of the binding energy of excitonic trions in two-dimensional systems has shown that this energy is ten times greater than in three-dimensional systems [14]. This shows that the influence of confinement results in enhanced binding energy. As a result, excitonic trions are more readily detectable in low-dimensional systems than in bulk semiconductors . ($3D$)

The positive trionsX_2^+ have been observed by photoluminescence and magnetic field photoreflectance experiments in Silicon acceptor-doped GaAs/AlGaAs quantum wells [18,19]. These experiments showed that increasing temperature facilitates thermal dissociation $X_2^+ \rightarrow X + h$, and that the binding energy of the second hole in the positive trion is only slightly different from that corresponding to the energy of the second electron in the negative trion. This contrasts with studies carried out by [14] in the case of two-dimensional semiconductors, where the difference in binding energy is over 30%.

In this chapter, we use the variational method to calculate the energy of negative excitonic trions for two cases, finite and infinite wells, resulting from coulombic coupling between two electrons and a hole, in a   semiconductor QB of cylindrical geometry composed of . $GaAs/AlGaAs$

II-Variational calculation of negative trion energies

A-Case of a finite well

1-Effective system Hamiltonian

We study the negative trion X^- formed by two electrons and one hole. This type of trion, which can be assimilated to the positive trion X^+ by exchanging electrons and holes, is confined in a cylindrical BQ of finite potential, height H and radius R with a base of semiconductor constituted by/$Ga_{1-x}Al_xAs$.

In the effective mass approximation and in a model with two simple non-degenerate parabolic bands, the Hamiltonian is written:

$$H_{trion}(r_e, r_h, P, T) = -\frac{\hbar^2}{2m_{e1}^*}\nabla_{e1}^2 - \frac{\hbar^2}{2m_{e2}^*}\nabla_{e2}^2 - \frac{\hbar^2}{2m_h^*}\nabla_h^2 + \frac{e^2}{\varepsilon_0(T)}\left[\frac{1}{\sqrt{(r_{e1}-r_{e2})^2+(z_{e1}-z_{e2})^2}} - \right.$$

$$\left. \frac{1}{\sqrt{(r_{e1}-r_h)^2+(z_{e1}-z_h)^2}} - \frac{1}{\sqrt{(r_{e2}-r_h)^2+(z_{e2}-z_h)^2}}\right] + V_\omega^{e1}(r_{e1}, z_{e1}, T) +$$

$$V_\omega^h(\rho_h, z_h, T) + V_\omega^{e2}(r_{e2}, z_{e2}, T),$$

$$(3.1)$$

with :

$$V_\omega^i(r_i z_i, P, T) = \begin{cases} 0 & si \ \rho_i \le R \ et \ |Z_i| \le d \\ V_i(P,T) & sinon \end{cases} \quad (i = e_1, e_2, h) \ V_{e1} = V_{e2}, \tag{3.2}$$

V_e et V_h Are the heights of the confinement potential barriers for the electron and the hole.

The electrostatic potential and distance origins are taken from the middle of the BQ. $m_{e,1-2}^*$ and m_h^* are the effective masses of the electrons and hole respectively, assumed constant in the box and in the barriers, and ε_0 is the dielectric constant of the material.

The temperature-dependent dielectric constant and effective masses of the electrons and hole are given by the following relationships [18-21] :

$$\varepsilon_0(T) = \begin{cases} 12.7 \exp\big(9.4 \times 10^{-5}(T - 75.6)\big) & for \ 0 \le T \le 200 \ K \\ 13.8 \exp\big(20.4 \times 10^{-5}(T - 300)\big) & for \ T > 200 \ K \end{cases} \tag{3.3}$$

and

$$\begin{cases} m_e^*(T) = m_0 \left[1 + 7.51 \left(\frac{2}{E_g(T)} + \frac{1}{E_g(T)+\Delta_0}\right)\right]^{-1} \\ m_h^*(T) = m_0(0.09 - 3.55 \times 10^{-5}T) \end{cases} \tag{3.4}$$

with m_0 is the mass of the free electron and Δ_0 is the orbital spin

In the following, we use atomic units, a_e^* for length and $2R_e^*$ for energy, which are defined by :

$$\sigma = \frac{m_e^*}{m_h^*} \qquad\qquad a_e^* = \frac{\varepsilon_0 \hbar^2}{m_e^* e^2} \qquad\qquad 2R_{ex}^* = \frac{m_e^* e^4}{\varepsilon_0^2 \hbar^2}$$

$$\tag{3..5}$$

Equation (3.1) then becomes :

$$H_{trion}(r_e, r_h, P, T) = -\frac{m_{e1}^*}{m_{e1}^*(T)} \nabla_{e1}^2 - \frac{m_{e2}^*}{m_{e2}^*(T)} \nabla_{e2}^2 - \frac{\sigma m_h^*}{m_h^*(T)} \nabla_h^2 +$$

$$\frac{\varepsilon_0(0)}{\varepsilon_0(T)} \left[\frac{1}{\sqrt{(r_{e1}-r_{e2})^2+(z_{e1}-z_{e2})^2}} - \frac{1}{\sqrt{(r_{e1}-r_h)^2+(z_{e1}-z_h)^2}} - \right.$$

$$\left. \frac{1}{\sqrt{(r_{e2}-r_h)^2+(z_{e2}-z_h)^2}}\right] + V_\omega^{e1}(r_{e1}, z_{e1}, T) + V_\omega^h(\rho_h, z_h, T) + V_\omega^{e2}(r_{e2}, z_{e2}, T),$$

$$\tag{3..6}$$

where:

$$V_w^i(r_i, z_i, T) = V_{wr}^i(r_i, T) + V_{wz}^i(z_i, T) - \delta V_i(r_i, z_i, P, T) \quad (\,i = e_1, e_2, h) \tag{3.7}$$

where V_{wp}^i and V_{wz}^i are the lateral and axial confinement potentials respectively, and δV_i is a corrective term. These potentials are expressed by :

$$\begin{cases} V^i_{wz}(\rho_i, P, T) = V_i\,\theta(\rho_i - R) \\ V^i_{\omega z}(z_i, P, T) = V_i\theta\left(|z_i| - \frac{H}{2}\right) \end{cases} ; \quad (i = e_1, e_2, h), \qquad (3.8)$$

where θ is the Heaviside function

$$\delta V_i = \begin{cases} 0 & si \ \theta_i < R \ ou \ |z_i| < H/2 \\ V_i & ailleurs \end{cases} (i = e_1, e_2, h). \qquad (3.9)$$

2- Choice of test function

The preceding Schrödinger equation cannot be solved exactly, so we'll have to make do with an approximate method. We have chosen to determine the ground state using the Ritz variational method. This consists in minimizing the average total energy with respect to the variational parameters involved in the chosen test wave function:

$$\psi_{trion}(r_{e1}, r_{e2}, r_h, z_{e1}, z_{e2}, z_h, T) =$$

$$F_{e1}(r_{e1}, z_{e1}, T)F_{e1}(r_{e1}, z_{e1}, T)F_h(r_h, z_h, T)\emptyset(r_{e1}, r_{e2}, r_h, z_{e1}, z_{e2}, z_h, T),$$

$$(3.10)$$

where$\emptyset$ is the wave function describing the internal motion of the trions defined by :

$$\emptyset(r_{e1}, r_{e2}, r_h, z_{e1}, z_{e2}, z_h, T) = exp[(-\alpha_1|r_{e1} - r_{e2}| - \alpha_2|r_{e1} - r_h| - \alpha_3|r_{e2} - r_h|) -$$

$$\gamma_1(z_{e1} - z_{e2})^2 - \gamma_2(z_{e1} - z_h)^2 - \gamma_3(z_{e2} - z_h)^2] \qquad (3.11)$$

$$F_i(r_i, z_i, T) = f_i(r_i, T)g_i(z_i, T) \qquad (3.12)$$

$\alpha_{i,1-3}$ and$\gamma_{i,1-3}$ are variational parameters. The form of the wave function given equation (3.12) not only satisfies the requirements of the strong-interaction region (as in very narrow BQs), but also gives the correct results near the global boundaries (weak-interaction region).

where$f_i(\rho_i, T)$ and$g_i(, z_i, T)$ are obtained by solving the two Schrödinger equations corresponding to the following two dimensions, lateral and axial:

$$\begin{cases} \left[-\frac{\hbar^2}{2m^*_i(T)}\nabla^2_i + V^i_\omega(r_i, T)\right]f_i(r_i, T) = E_i(r_i, T)f_i(r_i, T) \\ \left[-\frac{\hbar^2}{2m^*_i(T)}\nabla^2_i + V^i_\omega(z_i, T)\right]g_i(z_i, T) = E_i(z_i, T)g_i(z_i, T) \end{cases} (i = e_1, e_2, h) \qquad (3.13)$$

The solutions of this system of differential equations are of the following form :

$$f_i(r_i, T) = \begin{cases} J_0\left(\theta_i \frac{r_i}{R}\right) & pour & r_i \leq R \\ A_i K_0(\beta_i r_i) & pour & r_i \geq R \end{cases} \quad (i = e_1, e_2, h).$$

$$(3.14)$$

$$g_i(z_i, T)_{i=} \begin{cases} \cos\left(\pi_i \frac{z_i}{H}\right) & si & |z_i| < \frac{H}{2} \\ B_i \exp(-K_i |z_i|) & si & |z_i| \geq \frac{H}{2} \end{cases} \quad (i = e_1, e_2, h).$$

$$(3.15)$$

J_0 and K_0 are modified Bessel functions of order 0. The parameters $, \theta_i(T)\pi_i(T), A_i(T)$ and $B_i(T)$ are determined from the boundary conditionsà $r_i = R$ and $|z_i| = \frac{H}{2}$.

This allows us to write :

$$A_i = \frac{J_0(\theta_i)}{K_0(\beta_i R)} \quad \frac{\theta_i J_1(\theta_i)}{J_0(\theta_i)} = \frac{\beta_i R K_i(\beta_i R)}{K_0(\beta_i R)} \quad . \tag{3.16}$$

$$B_i = \cos\left(\frac{\pi_i}{2}\right) / \exp\left(-\frac{K_i H}{2}\right) \tan\left(\frac{\pi_i}{2}\right) = K_i\left(\frac{H}{\pi_i}\right) \quad . \tag{3.17}$$

$$\text{with} \qquad K_i = \sqrt{\frac{v_i}{\tau_i} - \left(\frac{\pi_i}{H}\right)^2} , \beta_i = \sqrt{\frac{v_i}{\tau_i} - \left(\frac{\theta_i}{R}\right)^2} .$$

$$(3.18)$$

In the atomic unit system: $\tau_{e1,e2} = 1$ and $\tau_h = \sigma = \frac{m^*_{e,1-2}}{m^*_h}$

3- Binding energy of excitonic trions

To describe the relative motion of the three particles, we can use cylindrical coordinates, defining the triangle e1e2h (Fig. (3.1)). The orientation of the triangle in space is specified by the cylindrical coordinates r_i, z_i et φ_i (i=1, 2, h), noting the rotation around the oz axis. For reasons of symmetry, the ground state is invariant to rotation about the oz axis, so we can ignore the φ variable.

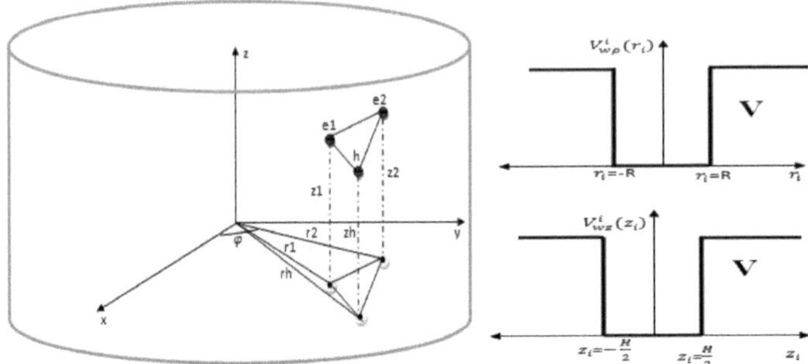

Figure 3.1: Schematic representation of the cylindrical QB and the containment model considered in this work.

The wave function and energy of the excitonic trion in the ground state are determined by minimizing the average energy:

$$E_{trion}(T)= \min_{\alpha_i \gamma_i}\langle \psi_{trion}|H_{eff}|\psi_{trion}\rangle \qquad (i=1,2,3).$$

(3.19)

$$\langle E(\alpha_i,\gamma_i)\rangle = \frac{\langle \psi_{trion}|\widehat{H}|\psi_{trion}\rangle}{\langle \psi_{trion}|\psi_{trion}\rangle}, \qquad (3.20)$$

the integral form of the dominator of equation (3.20) is:

$$\langle \psi_{trion}|H_{eff}|\psi_{trion}\rangle=$$

$$\int_{-\infty}^{+\infty}\int_{-\infty}^{+\infty}\int_{-\infty}^{+\infty}\int_{0}^{+\infty}\int_{0}^{+\infty}\int_{0}^{+\infty} \Psi_{trion}{}^*H_{eff}\,\Psi_{trion}\,r_{e1}dr_{e1}\,r_{e2}dr_{e2}r_h dr_h\,dz_{e1}dz_{e2}dz_h.$$

(3.21)

In our calculation we use the following mathematical formula, which is demonstrated in Appendix B:

$$-\int_{0}^{+\infty}\Psi^*\left(\frac{d^2}{dr^2}+\frac{1}{r}\frac{d}{dr}\right)\Psi\,r\,dr = \int_{0}^{+\infty}\left(\frac{d\Psi}{dr}\right)^2 r\,dr.$$

(3.22)

The form of the integral of the denominator of equation (3.20) is written :

$$\int_{-\infty}^{+\infty}\int_{-\infty}^{+\infty}\int_{-\infty}^{+\infty}\int_{0}^{+\infty}\int_{0}^{+\infty}\int_{0}^{+\infty} \Psi_{trion}{}^2\,r_{e1}\,dr_{e1}\,r_{e2}dr_{e2}r_h dr_h\,dz_{e1}dz_{e2}dz_h.$$

(3.23)

So

$$\langle E(\alpha_i,\gamma_i)\rangle =$$

$$\frac{\langle \psi_{trion}|\widehat{H}|\psi_{trion}\rangle}{\langle \psi_{trion}|\psi_{trion}\rangle}-\frac{m^*_{e1,2}}{m^*_{e1,2}(T)}\left[\frac{P_2-2P_4+P_6+P_{2A}+2P_{4A}+P_{6A}}{P_1}\right]+\frac{m^*_h}{m^*_h(T)}\sigma\left[\frac{P_{2B}+2P_{4B}+P_{6B}}{P_1}\right]+$$

$$\frac{m_{e1}^*}{m_{e1}^*(T)}\left[2(\gamma_1+\gamma_2)+\left(\frac{\pi_{e1}}{H}\right)^2+4\left(\frac{Z_3-Z_2}{Z_1}\right)\right]+\frac{m_{e2}^*}{m_{e2}^*(T)}\left[2(\gamma_1+\gamma_3)+\left(\frac{\pi_{e2}}{H}\right)^2-$$

$$4\left(\frac{Z_{3A}+Z_{2A}}{Z_1}\right)\right]+\frac{m_h^*}{m_h^*(T)}\sigma\left[2(\gamma_2+\gamma_3)+\left(\frac{\pi_h}{H}\right)^2-4\left(\frac{Z_{3B}+Z_{2B}}{Z_1}\right)\right]+V_{e1}\left(1-\frac{P_7}{P_1}\right)\times\frac{Z_5}{Z_1}$$

$$+V_{e2}\left(1-\frac{P_{7A}}{P_1}\right)\times\frac{Z_{5A}}{Z_1}+V_h\left(1-\frac{P_{7B}}{P_1}\right)\times\frac{Z_{5B}}{Z_1}+$$

$$\frac{\varepsilon_0(0)}{\varepsilon_0(T)}\left[\frac{Z_C}{Z_1}-\frac{Z_{CA}}{Z_1}-\frac{Z_{CB}}{Z_1}\right]. \tag{3.24}$$

The integrals P_i and Z_i are defined in Appendix B.

The temperature-dependent binding energy of the trion is defined as follows:

$$E_B(T)=E_i(T)+E_{exe}(T)-E_{trion}(T), \tag{3.25}$$

where $E_i(T)$ is the ground state energy of the uncorrelated electron e1, e2 and hole [20]. $E_{exe}(T)$ is the ground state energy of the exciton in the quantum dot as presented by equation (2.26) in chapter 2, and $E_{trion}(T)$ is the ground state energy of the negative trion calculated from equation (3.24).

B-Infinite well case

1-Effective Hamiltonian of the system.

Let's consider a negative trion, confined in a QB of cylindrical geometry of $GaAs$ / $AlAs$ of dimension (R,H) . Within the framework of the effective mass approximation and the envelope function, the Hamiltonian of such a system under the effect of temperature T considering an infinite confinement potential *Figure 3.2* can be written as follows:

$$H_{trion}(r_e,r_h,P,T)=-\frac{m_{e1}^*}{m_{e1}^*(T)}\nabla_{e1}^2-\frac{m_{e2}^*}{m_{e2}^*(T)}\nabla_{e2}^2-\frac{\sigma m_h^*}{m_h^*(T)}\nabla_h^2+$$

$$\frac{\varepsilon_0(0)}{\varepsilon_0(T)}\left[\frac{1}{\sqrt{(r_{e1}-r_{e2})^2+(z_{e1}-z_{e2})^2}}-\frac{1}{\sqrt{(r_{e1}-r_h)^2+(z_{e1}-z_h)^2}}-\right.$$

$$\left.\frac{1}{\sqrt{(r_{e2}-r_h)^2+(z_{e2}-z_h)^2}}\right]+V_\omega^{e1}(r_{e1},z_{e1},T)+V_\omega^h(\rho_h,z_h,T)+V_\omega^{e2}(r_{e2},z_{e2},T)$$

$$\tag{3.26}$$

$$\text{with: } (\quad V_\omega^i(r_iz_i,P,T)=\begin{cases}0 & si\ \ \rho_i\leq R\ et\ |Z_i|\leq d\\ \infty & sinon\end{cases}\qquad i=e_1,e_2,h) \tag{3.27}$$

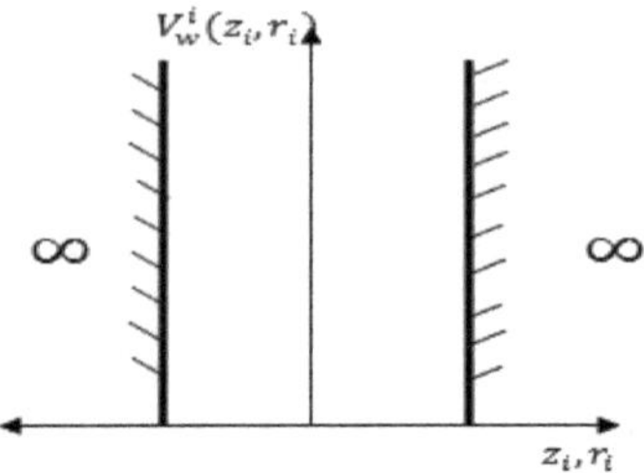

Figure 3.2: shows the radial and axial infinite potential well.

2-Test function

It is natural here to adopt a wave function of the same form as in the finite potential well approximation as follows:

$$\psi_{trion}(r_{e1}, r_{e2}, r_h, z_{e1}, z_{e2}, z_h, T) =$$
$$F_{e1}(r_{e1}, z_{e1}, T)F_{e1}(r_{e1}, z_{e1}, T)F_h(r_h, z_h, T)\emptyset(r_{e1}, r_{e2}, r_h, z_{e1}, z_{e2}, z_h, T)$$
(3.28)

where $\emptyset$ is the wave function describing the internal motion of the trions is defined in equation (3.11):

However, this function is separable into r_i and z_i $(i = 1,2,3)$, and leads to cumbersome calculations containing six variational parameters. We have therefore chosen a simpler non-separable wave function. This is a numerical evaluation of sextuplet integrals:

$$\psi_{trion}(r_{e1}, r_{e2}, r_h, z_{e1}, z_{e2}, z_h, T) =$$
$$F_{e1}(r_{e1}, z_{e1}, T)F_{e1}(r_{e1}, z_{e1}, T)F_h(r_h, z_h, T)\chi(r_{e1}, r_{e2}, r_h, z_{e1}, z_{e2}, z_h, T)$$
(3.29)

$$\chi(r_{e1}, r_{e2}, r_h, z_{e1}, z_{e2}, z_h, T) = exp\left[\left(-\alpha_1\sqrt{(r_{e1} - r_{e2})^2 + (z_{e1} - z_{e2})^2} - \right.\right.$$
$$\left.\left.\alpha_2\sqrt{(r_{e1} - r_h)^2 + (z_{e1} - z_h)^2} - \alpha_3\sqrt{(r_{e2} - r_h)^2 + (z_{e2} - z_h)^2}\right)\right], \qquad (3.30)$$

three variational parameters α_1, α_2 and α_3 are introduced. In this form, the wave function leads to simpler numerical calculations (no time problem during numerical calculation and the six variational parameters are limited to three).

$$F_i(r_i, z_i, T) = f_i(r_i, T)g_i(z_i, T) \quad (i=1, 2, h), (3.31)$$

where$f_i(\rho_i, T)$ and$g_i(, z_i, T)$ are obtained by solving the two Schrödinger equations corresponding to the following lateral and axial dimensions:

$$\begin{cases} \left[-\dfrac{\hbar^2}{2m_i^*(T)} \nabla_i^2 + V_\omega^i(r_i, T) \right] f_i(r_i, T) = E_i(r_i, T) f_i(r_i, T) \\ \left[-\dfrac{\hbar^2}{2m_i^*(T)} \nabla_i^2 + V_\omega^i(z_i, T) \right] g_i(z_i, T) = E_i(z_i, T) g_i(z_i, T) \end{cases} \quad (i = e_1, e_2, h). \quad (3.32)$$

The solutions of this system of differential equations are of the following form :

$$f_i(r_i, T) = \begin{cases} J_0\left(\theta_i \dfrac{r_i}{R}\right) & pour & r_i \leq R \\ 0 & pour & r_i \geq R \end{cases} \quad (i = e_1, e_2, h), \quad (3.33)$$

$$g_i(z_i, T)_{i=} \begin{cases} \cos\left(\pi_i \dfrac{z_i}{H}\right) & si & |z_i| < \dfrac{H}{2} \\ 0 & si & |z_i| \geq \dfrac{H}{2} \end{cases} \quad (i = e_1, e_2, h), \quad (3.34)$$

whereJ_0 is the Bessel function of order 0 with the first zero the value of$\theta_0 = 2.404\,8255577$.

3-Binding energy of excitonic trions

The equation, at eigenvalue $H\Psi = E\Psi$ where the Hamiltonian H is given by equation (3.26), is solved analytically and numerically to obtain the lowest energy values taking into account the geometric confinement in a quantum dot based on $GaAs/AlAs$. The binding energy of the negatively charged exciton is defined as :

$$E_B(T) = 2E_e(T) + E_h(T) + E_{exe}(T) - E_{trion}(T), \qquad .$$

(3.35)

with$E_{exe}(T)$ is the ground state energy of the exciton.$E_{e(h)}(T)$ is the ground state energy of the electron (hole) without Coulomb interaction.$E_{trion}(T)$ is the corresponding energy eigenvalue of the Hamiltonian. The latter is calculated by minimizing the variationnele parameters of the wave function:

$$\langle E(\alpha_i) \rangle = \frac{\langle \psi_{trion} | \hat{H} | \psi_{trion} \rangle}{\langle \psi_{trion} | \psi_{trion} \rangle}. \qquad . (3.36)$$

In the context of the infinite confinement potential considered in this section, we have :

$\langle \Psi_{trion} | \Psi_{trion} \rangle =$

$$\int_0^R \int_0^R \int_0^R \int_{-\frac{H}{2}}^{+\frac{H}{2}} \int_{-\frac{H}{2}}^{+\frac{H}{2}} \int_{-\frac{H}{2}}^{+\frac{H}{2}} f_{e1}{}^2 f_{e2}^2 f_h^2 g_{e1}^2 g_{e2}^2 g_h^2 \chi^2 \, r_{e1} \, dr_{e1} \, r_{e2} dr_{e2} r_h dr_h \, dz_{e1} dz_{e2} dz_h \qquad .$$

(3.37)

And

$$\langle\psi_{trion}|\widehat{H}|\psi_{trion}\rangle=$$

$$\frac{m_{e1}^*}{m_{e1}^*(T)}\left[\int_0^R\int_0^R\int_0^R\int_{-\frac{H}{2}}^{+\frac{H}{2}}\int_{-\frac{H}{2}}^{+\frac{H}{2}}\int_{-\frac{H}{2}}^{+\frac{H}{2}}f_{e2}^2f_h^2g_{e1}^2g_{e2}^2g_h^2\left(\frac{d}{dr_{e1}}(f_{e1}\chi)\right)^2 r_{e1}\,dr_{e1}\,r_{e2}dr_{e2}r_hdr_h\,dz_{e1}dz_{e2}dz_h\,-$$

$$\int_0^R\int_0^R\int_0^R\int_{-\frac{H}{2}}^{+\frac{H}{2}}\int_{-\frac{H}{2}}^{+\frac{H}{2}}\int_{-\frac{H}{2}}^{+\frac{H}{2}}f_{e1}^2f_{e2}^2f_h^2g_{e2}^2g_h^2\frac{d^2}{dz_{e1}^2}(g_{e1}\chi)\,r_{e1}\,dr_{e1}\,r_{e2}dr_{e2}r_hdr_h\,dz_{e1}dz_{e2}dz_h\,\right]\qquad +$$

$$\frac{m_{e2}^*}{m_{e2}^*(T)}\left[\int_0^R\int_0^R\int_0^R\int_{-\frac{H}{2}}^{+\frac{H}{2}}\int_{-\frac{H}{2}}^{+\frac{H}{2}}\int_{-\frac{H}{2}}^{+\frac{H}{2}}f_{e1}^2f_h^2g_{e1}^2g_{e2}^2g_h^2\left(\frac{d}{dr_{e2}}(f_{e2}\chi)\right)^2 r_{e1}\,dr_{e1}\,r_{e2}dr_{e2}r_hdr_h\,dz_{e1}dz_{e2}dz_h\,-$$

$$\int_0^R\int_0^R\int_0^R\int_{-\frac{H}{2}}^{+\frac{H}{2}}\int_{-\frac{H}{2}}^{+\frac{H}{2}}\int_{-\frac{H}{2}}^{+\frac{H}{2}}f_{e1}^2f_h^2f_{e2}^2g_{e1}^2g_h^2\frac{d^2}{dz_{e2}^2}(g_{e2}\chi)\,r_{e1}\,dr_{e1}\,r_{e2}dr_{e2}r_hdr_h\,dz_{e1}dz_{e2}dz_h\,\right]+$$

$$\frac{m_h^*\sigma}{m_h^*(T)}\left[\int_0^R\int_0^R\int_0^R\int_{-\frac{H}{2}}^{+\frac{H}{2}}\int_{-\frac{H}{2}}^{+\frac{H}{2}}\int_{-\frac{H}{2}}^{+\frac{H}{2}}f_{e1}^2f_{e2}^2g_{e1}^2g_{e2}^2g_h^2\left(\frac{d}{dr_h}(f_h\chi)\right)^2 r_{e1}\,dr_{e1}\,r_{e2}dr_{e2}r_hdr_h\,dz_{e1}dz_{e2}dz_h\,-$$

$$\int_0^R\int_0^R\int_0^R\int_{-\frac{H}{2}}^{+\frac{H}{2}}\int_{-\frac{H}{2}}^{+\frac{H}{2}}\int_{-\frac{H}{2}}^{+\frac{H}{2}}f_{e1}^2f_{e2}^2f_h^2g_{e1}^2g_{e2}^2\frac{d^2}{dz_h^2}(g_h\chi)\,r_{e1}\,dr_{e1}\,r_{e2}dr_{e2}r_hdr_h\,dz_{e1}dz_{e2}dz_h\,\right]+$$

$$\frac{\varepsilon_0(0)}{\varepsilon_0(T)}\left[\int_0^R\int_0^R\int_0^R\int_{-\frac{H}{2}}^{+\frac{H}{2}}\int_{-\frac{H}{2}}^{+\frac{H}{2}}\int_{-\frac{H}{2}}^{+\frac{H}{2}}f_{e2}^2f_h^2g_{e1}^2g_{e2}^2g_h^2(f_{e1}\chi)^2\left[\frac{1}{\sqrt{(r_{e1}-r_{e2})^2+(z_{e1}-z_{e2})^2}}\,-\right.\right.$$

$$\left.\left.\frac{1}{\sqrt{(r_{e1}-r_h)^2+(z_{e1}-z_h)^2}}-\frac{1}{\sqrt{(r_{e2}-r_h)^2+(z_{e2}-z_h)^2}}\right] r_{e1}\,dr_{e1}\,r_{e2}dr_{e2}r_hdr_h\,dz_{e1}dz_{e2}dz_h\,\right]\qquad .$$

(3.38)

III-Numerical results and discussion

In this section, we present our numerical results based on the theoretical approach outlined in the previous section. As already mentioned, the BQ material studied is based on $GaAs$. The height of the confinement potential is induced by the barrier material based on $Ga_{1-x}Al_xAs$. The method of numerical resolution is still based (chap.2) on the minimization of the energy with respect to the variational parameters.

The effective electron and hole masses corresponding to $GaAs$ are respectively: $m_e^* = 0.063\,m_0$ et $m_h^*=0.079\,m_0$. Our results are expressed in atomic units, the effective Rydberg $R_e^* = 4.8438\,meV$ and the effective Bohr radius $a_e^* = 11.172\,nm$.

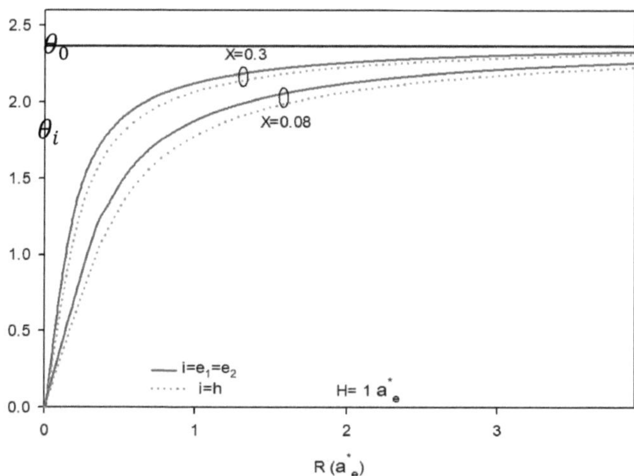

Figure 3.3: *Variation of wave function parameters θi (i=e$_{1,2}$ (solid lines, i=h (dashed lines)) as a function of cylinder radius R for two values of fraction x ofAl in .Ga$_{1-x}$Al$_x$As*

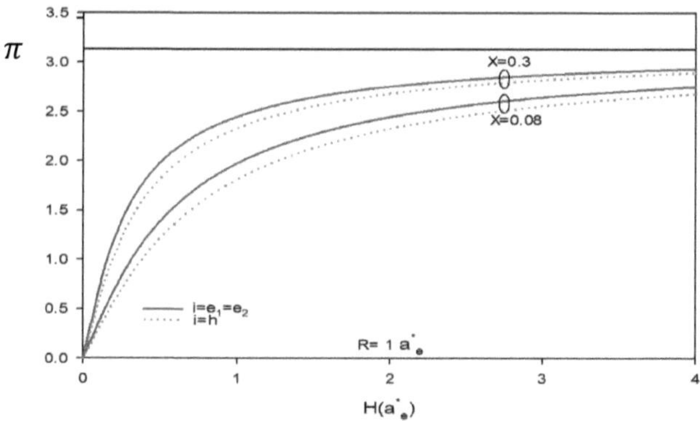

Figure 3.4: *Variation of the parametersπ i (i=e$_{1,2}$ (solid lines, i=h (dashed lines)) of the wave function as a function of cylinder height H for two values of the fraction x ofAl in*

$$.Ga_{1-x}Al_xAs$$

In order to verify that the infinite confinement potential remains inadequate to describe the excitonic effect in the case of thin QBs, and to present the interest of the finite confinement chosen here, we illustrate in *Figure 3.3* the variation of the θ_e (electron case) andθ_h (hole case) parameters of the wave function as a function of disk radius for two aluminum mole fractions ($x = 0.08$ and 0.3). For each value of x, the parameters characterizing the lateral wave function undergo an increase with radius R, reaching the characteristic value of the infinite potential for the limit of large values of R. For lateral confinement, at each box radius R, the parameter values become significant for large concentrations $x=0.3$. For axial confinement, **Figure 3.4** shows the variation of the parametersπ_e (electron case) and π_h (hole case) as a function of cylinder height H for both fractions ($x = 0.08$ and 0.3). The curve obtained behaves in a similar way to the previous one, showing that the wave function parameters characterizing axial confinement increase considerably with height until they reach the characteristic value for infinite confinement. From the above, it is therefore interesting to consider the case of a finite potential

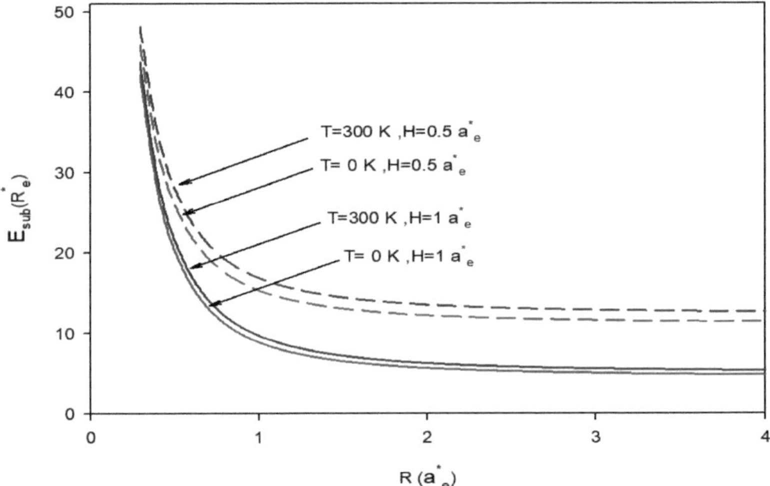

Figure 3.5: Variation of subband energy as a function of QB radius for different values of H and temperature.

To illustrate these results, **Figure 3.5 shows** the sub-band energy as a function of radius for different temperatures ($T = 0$ *and* 300 K) and different H heights (H=0.5 1 a^*_e et a^*_e). The aluminum mole fraction is set at $x = 0.3$. It appears that for each value of T, the energy$E_{sub} \to \infty$ when $R \to 0$. For fixed R, E_{sub} is an increasing function of T, and when $R \to \infty$, we obtain the quantum well limit.

In ***Figure 3.6,*** the subband energy shows the same behavior as a function of the height H relative to the radius R; it decreases with increasing box height These results are qualitatively in agreement with those obtained in the case of the effect of size on exciton states in the cylindrical QB with a classical confinement model [21]. Note that effect of temperature is of the same nature for large QB heights.

Furthermore, for each temperature value, we see a decrease in the size of the box, which leads to an increase in the subband energy of the trion. This means that the charged exciton remains stable even at room temperature. This behavior justifies the fact that size effects induce extra-confinement in addition to the potential confinement adopted. It's also important to note that, for each value of size, the sub-band energy increases with increasing T, which tends to decrease the electron-hole attraction, and consequently the charged exciton loses its stability. The practical interest of these opposing effects can be exploited to improve the optical properties linked to the existence of trions at different temperatures, while maintaining the size of the QB.

To give a clear picture of the behavior of the subband energy at different temperatures, we have plotted in ***Figure 3.7*** the energy of the negatively charged exciton subband as a function of *aluminum* concentration x for a radius $R=1a_e^*$ and a height $H=1a_e^*$. It appears that the subband energy is strongly dependent on the mole fraction x in the barrier material. For each value of temperature T, the sub-band energy follows an almost linear variation with the concentration x, and this variation becomes significant for higher values of temperature. It is also important to note that the excitonic wave function extends easily outwards from the QB for barrier materials weakly doped with aluminum. As shown in the previous figure, the sub-band energy is also important in the case of small QB sizes, favouring trion stability and therefore the possibility of detection, which is non-existent in bulk semiconductors.

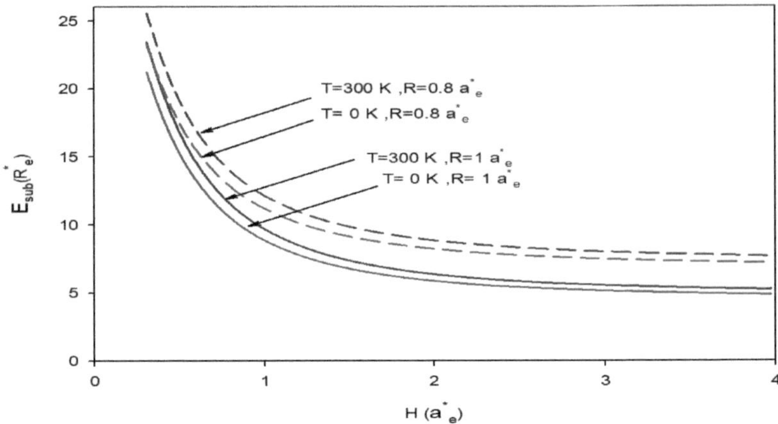

Figure 3.6: *Variation of sub-band energy as a function of cylinder height for different values of radius R and temperature T.*

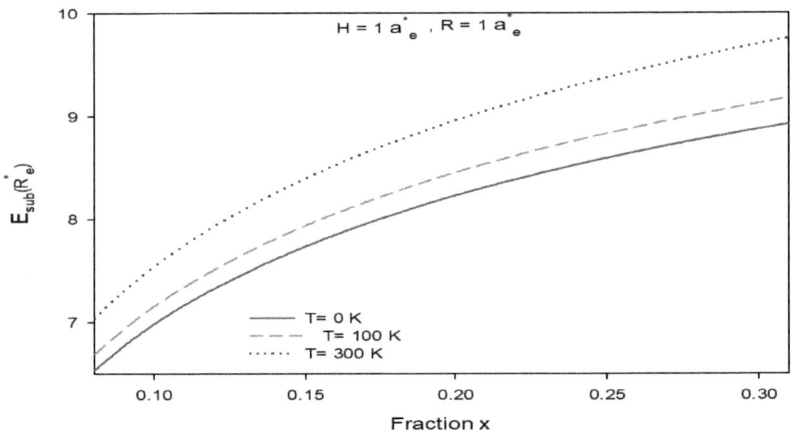

Figure 3.7: *Variation of subband energy as a function of mole fraction x , for different temperatures$(T = 0K, 100K, 300K)$ withH etR fixées   at 1 .a_e^**

The following two figures show the subband energy of a negatively charged exciton as a function of temperature and QB size, in the case of an infinite well.

To study the influence of QB radius and temperature T on electron1-electron2-hole non-correlation in our infinite-potential confinement model, we have illustrated in ***Figure 3.8*** the evolution of uncorrelated energy as a function of box radius for different values of height H and temperature. We use the classical definition of the sub-band energy given by the sum of the electron and hole energies. Clearly, the subband energy increases as the radius decreases. In other words, the width of the well is reduced until the subband energy tends towards infinity. This behavior is due to the infinite confinement potential model considered. The subband energy undergoes a considerable increase with temperature variation, and this for every value of radius R.

Figure 3.9 shows the behavior of subband energy as a function of quantum cylinder height, for different values of radius R ($R\ 1 = a_e^*$, 0.5 a_e^*) and temperature T (0 and 300 K). We note that the energy of the sub-band increases sharply with decreasing height H, and becomes significant compared to that shown in the previous figure. For a fixed value of H, the sub-band energy increases as T increases

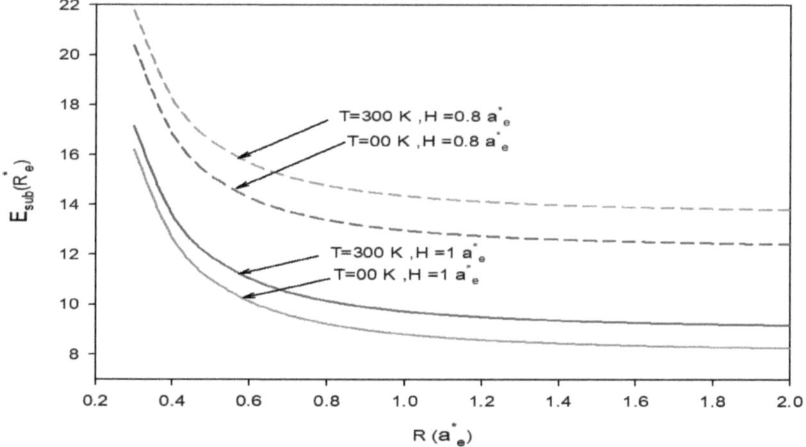

Figure 3.8: *Variation of sub-band energy as a function of cylinder radius R, for different temperatures* ($T = 0K, 300K$) *and different heights (H= $1a_e^*$,0.8a_e^*).*

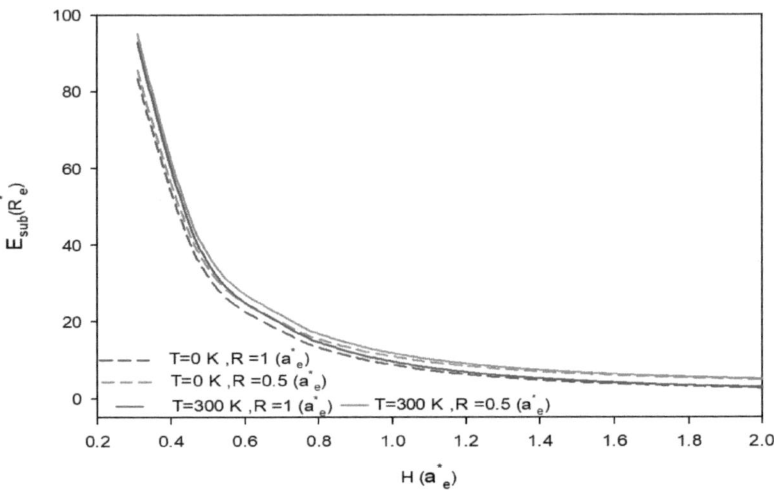

Figure 3.9: *Variation of sub-band energy as a function of cylinder height H, for different temperatures* $(T = 0K, 300K)$ *and radii* $(R = 1a_e^*, 0.5a_e^*)$.

Conclusion

In this chapter, we have evaluated the subband energy of a negatively charged exciton as a function of temperature in a cylindrical QB. The height of the potential barrier determined by the discontinuity of band gaps at the interface of the $GaAs/Ga_{1-x}Al_xAs$ heterostructure showed some temperature dependence. We also examined the effect of temperature on the non-correlation energy in the finite and infinite potential confinement model.

As far as the sub-band energy is concerned, by decreasing the can radius, this energy follows an increasing function up to infinity. In this respect, we have presented a discussion of its dependence on can dimensions, aluminum mole fraction x and temperature.

Our main contribution in this chapter was the effect of temperature on the non-correlated energy of the negative trion for two confinement models, finite and infinite. To our knowledge, this is the first study to evaluate this effect in a QB of cylindrical geometry, which also enabled us to qualitatively compare our results with those obtained by [21-22] dealing with the case of negative and positive trions. This is a preliminary step towards more complex studies, taking into account the effect of pressure, electric field and phonons.

Bibliography

[1]Lampert, Murray A. *Physical Review Letters* 1 (1958) 450.

[2]Wannier, Gregory H. *Physical Review* 52 (1937) 191.

[3]Haynes, J. R. *Physical Review Letters* 4 (1960) 361.

[4] Thomas, D. G., and J. J. Hopfield. *Physical Review* 128 (1962) 2135.

[5]D. C.Reynolds, C. W. Litton, and T. C. Collins. *Physical Review140* (1965) A1726.

[6]Dean, P. J., Fischer, B., Herbert, D. C., Lagois, J., & Yu, P. Y. . *Excitons"*. Vol. 14. Springer Science & Business Media, (2012).

[7]Pokutnyi, S. I. *Semiconductors* 47 (2013) 1626-1635.

[8]A. Guillaume, C. Benoît. *Phys. Rev.*177 (1969) 567.

[9]Haken, H., and S. Nikitine. "Excitons at high densities." *Springer Tracts Mod. Phys* 73 (1975).

[10]Davydov, A. "*Theory of molecular excitons*". Springer, (2013).

[11]Tsuchiya, Takuma. *Journal of luminescence* 87 (2000) 509-511.

[12]Munschy, G., and B. Stébé. *physica status solidi (b)* 64 (1974: 213-222.

[13]Teran, F. J., Eaves, L., Mansouri, L., Buhmann, H., Maude, D. K., Potemski, M. *Physical Review B* 71 (2005) 161309.

[14]Stébé, B., A. Ainane. *Superlattices and microstructures* 5 (1989) 545-548.

[15]Thilagam, A. *Physical Review B* 55 (1997) 7804.

[16] Thomas, G. A., and T. M. Rice. *Solid State Communications* 23 (1977) 359-363.

[17]B. Stébé,Thèsed'Etat,Metz (1977).

[18]Shields, A. J. Osborne, J. L., Simmons, M. Y., Pepper, M., and Ritchie, D. A. *Physical Review B52* (1995) R5523.

[19]Finkelstein, G., Shtrikman, H., and Bar-Joseph, I. *Physical Review B* 53 (1996) R1709.

[20]Wu, Shudong, and Li Wan. *Journal of Applied Physics111* (2012) 063711.

[21]Chafai, A., Essaoudi, I., Ainane, A., Dujardin, F., and Ahuja, R. *Physica E: Low-dimensional Systems and Nanostructures* 101 (2018) 125-130.

[22]Kezerashvili, R. Ya, and Kezerashvili, R. Y., Machavariani, Z. S., Beradze, B., *andTchelidzearXivpreprint arXiv: 1804.11030* (2018).

General conclusion

The aim of the work presented here was to determine the energy spectrum of negative excitons and trions as a function of QB dimensions, temperature and pressure. The exciton and trion studied are assumed to be confined within a cylindrical semiconductor-based QB. Throughout this work, the framework of approximations that we have deemed appropriate is that considering both the variational principle, the effective mass process and the notion of envelope function. We schematize quantum confinement as a finite and infinite potential well.

In the first chapter, we review the fundamentals of semiconductor physics and explain the various principles and approximation methods used to describe the electronic structures and optical properties of semiconductors.

In the second chapter, we began with the main features our theoretical model, which enabled us to determine the energy spectrum of the exciton confined in the cylindrical QB under the effect of pressure and temperature. In essence, we were able to evaluate the binding energy. This energy is highly dependent on the size of the box, the concentration of aluminium in the barrier material and the effect of pressure and temperature. The envelope function we have chosen has enabled us to examine the various boundary cases $R \rightarrow \infty, H \rightarrow \infty, R \rightarrow 0 \ et \ H \rightarrow 0$. In the latter two cases, the effective mass approximation loses its validity. We have also shown that the electron wave function is mainly confined to the central region and decreases as the size of the box increases, due to its tunneling into the barrier material. In connection with the optical properties, we have determined the integrated photoluminescence energy. This energy is sensitive to can size, temperature, pressure and the height of the potential barrier induced by the mole fraction of aluminum.

In the third chapter, we examined the effect of temperature negatively-charged exciton states in a cylindrical arsenic galium-based QB($GaAs$) embedded in the same barrier material doped with aluminum mole fractions. We found a high sensitivity of the non-correlation energy as a function of temperature variation, leading us to suggest through this energy a possible detection of excitonic trion lines at room temperature. This result is of great scientific and technological importance, since it enables us to operate at suitable temperature orders and even to take advantage of the properties of charged excitons in these orders and of the conditions to be chosen.

In the interest of examining the stability of excitonic trions in BQs as a function of temperature, also investigated in chapter 3 the influence of temperature on the energy spectrum of negative trions in an infinite confinement model. The energy of the subband was extensively discussed

It is highly desirable that these results should be exploited and experimented on, first to enrich them and then to give them their full scope. As a prospect, the study we have carried out in this work opens up a new avenue for other studies to be tackled concerning the effects of external fields on the energy spectrum of excitonic negative and positive trions, and biexcitons as well as their couplings with optical phonons.

Appendix A

This appendix contains details of the useful calculations relating to the cylindrical coordinate used in Chapters 2 and 3. It coincides with the expressions of the integral over the configuration space of the two particles of a function invariant by global rotation of the system around the cylinder axis.

Provided we consider only "A" operators involving the variables $\rho_1, z_1, \rho_2, z_2, \rho, \psi$ is a function of these 5 variables. The problem is then to show that the integral of a function $F(\rho_1, z_1, \rho_2, z_2, \rho,)$ when the double particles describe the whole space, can be put in the form of a quintuple integral:

$$\int_{-\infty}^{+\infty} dx_e \int_{-\infty}^{+\infty} dy_e \int_{-\infty}^{+\infty} dz_e \int_{-\infty}^{+\infty} dx_h \int_{-\infty}^{+\infty} dy_h \int_{-\infty}^{+\infty} dz_h F(\rho_e, z_e, \rho_h, z_h, \rho_{eh})$$

$$= \int_0^{+\infty} d\rho_e \int_{-\infty}^{+\infty} dz_1 \int_0^{+\infty} d\rho_h \int_{-\infty}^{+\infty} dz_2 \int_{|\rho_e-\rho_h|}^{(\rho_e-\rho_h)} d\rho_{eh} F(\rho_e, z_h, \rho_h, z_h, \rho_{eh}) W(\rho_1, z_e, \rho_2, z_h, \rho_{eh})$$

When we switch from Cartesian coordinates to cylindrical coordinates, a weighting function appears $W(\rho_e, z_e, \rho_h, z_h, \rho_{eh}) = \rho_e \rho_h$. Finally, the expression P_i becomes :

$$P_i = 8\pi \int_0^\infty d\rho_e \int_0^\infty d\rho_h \int_{|\rho_e-\rho_h|}^{\rho_e+\rho_h} \frac{F_i(\rho_e, \rho_h, \rho_{eh})\rho_e\rho_h\rho_{eh}}{\sqrt{[(\rho_e+\rho_h)^2 - \rho_{eh}^2][\rho_{eh}^2 - (\rho_e+\rho_h)^2]}} d\rho_{eh}$$

$$P_1 = 8\pi R^4[I_1 + A_e^2 I_2 + A_h^2 I_3 + A_e^2 A_h^2 I_4]$$
$$P_3 = 8\pi R^4[I_5 + A_e^2 I_6 + A_h^2 I_7 + A_e^2 A_h^2 I_8]$$
$$P_4 = -8\pi R^3 \frac{1}{(1+\sigma)}[\theta_e R I_9 + \beta_e A_e^2 R I_{10} + \theta_e A_h^2 I_{11} + \beta_e A_e^2 A_h^2 R I_{12}]$$
$$P_5 = -8\pi R^3 \frac{\sigma}{(1+\sigma)}[\theta_h R I_{13} + \theta_h A_e^2 R I_{14} + \beta_h A_h^2 I_{15} + \beta_h A_e^2 A_h^2 R I_{16}]$$
$$P_7 = 8\pi R^4[I_1 + A_h^2 I_3]$$

And

$$P_7 = 8\pi R^4[I_1 + A_e^2 I_2]$$

or

$$I_1 = \int_0^1 \int_0^1 \int_{R|x-y|}^{R(x+y)} xyu J_0^2(\theta_e x) J_0^2(\theta_h y) \frac{\exp(-2\alpha Ru)}{\{[(x+y)^2 - u^2][u^2 - (x-y)^2]\}} dxdydu$$

$$I_2 = \int_0^1 \int_0^1 \int_{R|x-y|}^{R(x+y)} \frac{1}{x^3} yu K_0^2(\beta_e \frac{R}{x}) J_0^2(\theta_h y) \frac{\exp(-2\alpha Ru)}{\left\{\left[\left(\frac{1}{x}+y\right)^2 - u^2\right]\left[u^2 - \left(\frac{1}{x}-y\right)^2\right]\right\}} dxdydu$$

$$I_3 = \int_0^1 \int_0^1 \int_{R|x-y|}^{R(x+y)} x \frac{1}{y^3} u J_0^2(\theta_e x) K_0^2(\beta_h \frac{R}{y}) \frac{\exp(-2\alpha Ru)}{\left\{\left[\left(x+\frac{1}{y}\right)^2 - u^2\right]\left[u^2 - \left(x-\frac{1}{y}\right)^2\right]\right\}} dxdydu$$

$$I_4 = \int_0^1 \int_0^1 \int_{R|x-y|}^{R(x+y)} \frac{1}{x^3} \frac{1}{y^3} u K_0^2(\beta_e \frac{R}{x}) K_0^2(\beta_h \frac{R}{y}) \frac{\exp(-2\alpha Ru)}{\left\{\left[\left(\frac{1}{x}+\frac{1}{y}\right)^2 - u^2\right]\left[u^2 - \left(\frac{1}{x}-\frac{1}{y}\right)^2\right]\right\}} dxdydu$$

$$I_5 = \int_0^1 \int_0^1 \int_{R|x-y|}^{R(x+y)} xy J_0^2(\theta_e x) J_0^2(\theta_h y) \frac{\exp(-2\alpha R u)}{\{[(x+y)^2 - u^2][u^2 - (x-y)^2]\}} dx\, dy\, du$$

$$I_6$$
$$= \int_0^1 \int_0^1 \int_{R|x-y|}^{R(x+y)} \frac{1}{x^3} y K_0^2\left(\beta_e \frac{R}{x}\right) J_0^2(\theta_h y) \frac{\exp(-2\alpha R u)}{\left\{\left[\left(\frac{1}{x}+y\right)^2 - u^2\right]\left[u^2 - \left(\frac{1}{x}-y\right)^2\right]\right\}} dx\, dy\, du$$

$$I_7$$
$$= \int_0^1 \int_0^1 \int_{R|x-y|}^{R(x+y)} x \frac{1}{y^3} J_0^2(\theta_e x) K_0^2\left(\beta_h \frac{R}{y}\right) \frac{\exp(-2\alpha R u)}{\left\{\left[\left(x+\frac{1}{y}\right)^2 - u^2\right]\left[u^2 - \left(x-\frac{1}{y}\right)^2\right]\right\}} dx\, dy\, du$$

$$I_8$$
$$= \int_0^1 \int_0^1 \int_{R|x-y|}^{R(x+y)} \frac{1}{x^3}\frac{1}{y^3} K_0^2\left(\beta_e \frac{R}{x}\right) K_0^2\left(\beta_h \frac{R}{y}\right) \frac{\exp(-2\alpha R u)}{\left\{\left[\left(\frac{1}{x}+\frac{1}{y}\right)^2 - u^2\right]\left[u^2 - \left(\frac{1}{x}-\frac{1}{y}\right)^2\right]\right\}} dx\, dy\, d$$

$$I_9$$
$$= \int_0^1 \int_0^1 \int_{R|x-y|}^{R(x+y)} y(u^2 + x^2$$
$$- y^2) J_0(\theta_e x) J_1(\theta_e x) J_0^2(\theta_h y) \frac{\exp(-2\alpha R u)}{\{[(x+y)^2 - u^2][u^2 - (x-y)^2]\}} dx\, dy\, du$$

$$I_{10}$$
$$= \int_0^1 \int_0^1 \int_{R|x-y|}^{R(x+y)} \frac{1}{x^2} y \left(u^2 + \frac{1}{x^2}\right.$$
$$\left. - y^2\right) K_0\left(\beta_e \frac{R}{x}\right) K_1\left(\beta_e \frac{R}{x}\right) J_0^2(\theta_h y) \frac{\exp(-2\alpha R u)}{\left\{\left[\left(\frac{1}{x}+y\right)^2 - u^2\right]\left[u^2 - \left(\frac{1}{x}-y\right)^2\right]\right\}} dx\, dy\, du$$

$$I_{11}$$
$$= \int_0^1 \int_0^1 \int_{R|x-y|}^{R(x+y)} x \frac{1}{y^3} \left(u^2 + x^2\right.$$
$$\left. - \frac{1}{y^2}\right) J_0(\theta_e x) J_1(\theta_e x) K_0\left(\beta_h \frac{R}{y}\right) \frac{\exp(-2\alpha R u)}{\left\{\left[\left(x+\frac{1}{y}\right)^2 - u^2\right]\left[u^2 - \left(x-\frac{1}{y}\right)^2\right]\right\}} dx\, dy\, du$$

$$I_{12}$$
$$= \int_0^1 \int_0^1 \int_{R|x-y|}^{R(x+y)} \frac{1}{x^2}\frac{1}{y^3} \left(u^2 + \frac{1}{x^2}\right.$$
$$\left. - \frac{1}{y^2}\right) K_0\left(\beta_e \frac{R}{x}\right) K_1\left(\beta_e \frac{R}{x}\right) K_0\left(\beta_h \frac{R}{y}\right) \frac{\exp(-2\alpha R u)}{\left\{\left[\left(\frac{1}{x}+\frac{1}{y}\right)^2 - u^2\right]\left[u^2 - \left(\frac{1}{x}-\frac{1}{y}\right)^2\right]\right\}} dx\, dy\, du$$

$$I_{13}$$
$$= \int_0^1 \int_0^1 \int_{R|x-y|}^{R(x+y)} x(u^2 + x^2$$
$$- y^2) J_0(\theta_e x) J_1(\theta_e x) J_0^2(\theta_h y) \frac{\exp(-2\alpha R u)}{\{[(x+y)^2 - u^2][u^2 - (x-y)^2]\}} dx\, dy\, du$$

I_{14}

$$= \int_0^1 \int_0^1 \int_{R|x-y|}^{R(x+y)} \frac{1}{x^3}\left(u^2 + y^2 \right.$$

$$\left. - \frac{1}{x^2}\right) K_0\left(\beta_e \frac{R}{x}\right) K_1\left(\beta_e \frac{R}{x}\right) J_0^2(\theta_h y) \frac{\exp(-2\alpha Ru)}{\left\{\left[\left(x+\frac{1}{y}\right)^2 - u^2\right]\left[u^2 - \left(x - \frac{1}{x}\right)^2\right]\right\}} \, dxdydu$$

I_{15}

$$= \int_0^1 \int_0^1 \int_{R|x-y|}^{R(x+y)} x\frac{1}{y^2}\left(u^2 + x^2 \right.$$

$$\left. - \frac{1}{y^2}\right) J_0^2(\theta_e x) k_1\left(\beta_h \frac{R}{y}\right) K_0\left(\beta_h \frac{R}{y}\right) \frac{\exp(-2\alpha Ru)}{\left\{\left[\left(x+\frac{1}{y}\right)^2 - u^2\right]\left[u^2 - \left(x - \frac{1}{y}\right)^2\right]\right\}} \, dxdydu$$

I_{16}

$$= \int_0^1 \int_0^1 \int_{R|x-y|}^{R(x+y)} \frac{1}{x^3}\frac{1}{y^2}\left(u^2 + \frac{1}{x^2} \right.$$

$$\left. - \frac{1}{y^2}\right) K_0^2\left(\beta_e \frac{R}{x}\right) K_0\left(\beta_h \frac{R}{y}\right) K_1\left(\beta_h \frac{R}{y}\right) \frac{\exp(-2\alpha Ru)}{\left\{\left[\left(\frac{1}{x}+\frac{1}{y}\right)^2 - u^2\right]\left[u^2 - \left(\frac{1}{x} - \frac{1}{y}\right)^2\right]\right\}} \, dxdydu$$

The same applies to the Zi integral. After developing its expression, we obtain :

$$Z_1 = \frac{H^2}{4}\left(B_e^2 J_1 + B_e^2 J_2 + B_h^2 J_3 + B_h^2 J_4 + B_e^2 B_h^2 J_5 + B_e^2 B_h^2 J_6 + B_e^2 B_h^2 J_7 + B_e^2 B_h^2 J_8 + J_9\right)$$

$$J_1 = \int_0^1 \int_{-1}^1 \frac{1}{x^2} exp\left(-2K_e \frac{H}{x}\right) \cos^2\left(\pi_h \frac{y}{2}\right) exp\left(-\frac{H^2}{2}\gamma\left(\frac{1}{x}+y\right)^2\right) dxdy$$

$$J_2 = \int_0^1 \int_{-1}^1 \frac{1}{x^2} exp\left(-2K_e \frac{H}{x}\right) \cos^2\left(\pi_h \frac{y}{2}\right) exp\left(-\frac{H^2}{2}\gamma\left(\frac{1}{x}-y\right)^2\right) dxdy$$

$$J_3 = \int_{-1}^1 \int_0^1 \frac{1}{y^2} \cos^2\left(\pi_e \frac{x}{2}\right) exp\left(-2K_h \frac{H}{y}\right) exp\left(-\frac{H^2}{2}\gamma\left(x+\frac{1}{y}\right)^2\right) dxdy$$

$$J_4 = \int_{-1}^1 \int_0^1 \frac{1}{y^2} \cos^2\left(\pi_e \frac{x}{2}\right) exp\left(-2K_h \frac{H}{y}\right) exp\left(-\frac{H^2}{2}\gamma\left(x-\frac{1}{y}\right)^2\right) dxdy$$

$$J_5 = \int_0^1 \int_0^1 \frac{1}{x^2}\frac{1}{y^2} exp\left(-2K_e \frac{H}{x}\right) exp\left(-2K_h \frac{H}{y}\right) exp\left(-\frac{H^2}{2}\gamma\left(\frac{1}{x}-\frac{1}{y}\right)^2\right) dxdy$$

$$J_6 = \int_0^1 \int_0^1 \frac{1}{x^2}\frac{1}{y^2} exp\left(-2K_e \frac{H}{x}\right) exp\left(-2K_h \frac{H}{y}\right) exp\left(-\frac{H^2}{2}\gamma\left(\frac{1}{x}-\frac{1}{y}\right)^2\right) dxdy$$

$$J_7 = \int_0^1 \int_0^1 \frac{1}{x^2}\frac{1}{y^2} exp\left(-2K_e \frac{H}{x}\right) exp\left(-2K_h \frac{H}{y}\right) exp\left(-\frac{H^2}{2}\gamma\left(\frac{1}{x}+\frac{1}{y}\right)^2\right) dxdy$$

$$J_8 = \int_0^1 \int_0^1 \frac{1}{x^2}\frac{1}{y^2} exp\left(-2K_e \frac{H}{x}\right) exp\left(-2K_h \frac{H}{y}\right) exp\left(-\frac{H^2}{2}\gamma\left(\frac{1}{x}+\frac{1}{y}\right)^2\right) dxdy$$

$$J_9 = \int_{-1}^1 \int_{-1}^1 \cos^2\left(\pi_e \frac{x}{2}\right) \cos^2\left(\pi_h \frac{y}{2}\right) exp\left(-\frac{H^2}{2}\gamma(x+y)^2\right) dxdy$$

$$Z_2 = \frac{H^4}{16}\left(B_e^2 J_{10} + B_e^2 J_{11} + B_h^2 J_{12} + B_h^2 J_{13} + B_e^2 B_h^2 J_{14} + B_e^2 B_h^2 J_{15} + B_e^2 B_h^2 J_{16} + B_e^2 B_h^2 J_{17} \right.$$

$$\left. + J_{18}\right)$$

$$J_{10} = \int_0^1 \int_{-1}^1 \frac{1}{x^2}\left(\frac{1}{x}+y\right)^2 exp\left(-2K_e\frac{H}{x}\right)\cos^2\left(\pi_h\frac{y}{2}\right)exp\left(-\frac{H^2}{2}\gamma\left(\frac{1}{x}+y\right)^2\right)dxdy$$

$$J_{11} = \int_0^1 \int_{-1}^1 \frac{1}{x^2}\left(\frac{1}{x}-y\right)^2 exp\left(-2K_e\frac{H}{x}\right)\cos^2\left(\pi_h\frac{y}{2}\right)exp\left(-\frac{H^2}{2}\gamma\left(\frac{1}{x}-y\right)^2\right)dxdy$$

$$J_{12} = \int_{-1}^1 \int_0^1 \frac{1}{y^2}\left(x+\frac{1}{y}\right)^2 \cos^2\left(\pi_e\frac{x}{2}\right)exp\left(-2K_h\frac{H}{y}\right)exp\left(-\frac{H^2}{2}\gamma\left(x+\frac{1}{y}\right)^2\right)dxdy$$

$$J_{13} = \int_{-1}^1 \int_0^1 \frac{1}{y^2}\left(x-\frac{1}{y}\right)^2 \cos^2\left(\pi_e\frac{x}{2}\right)exp\left(-2K_h\frac{H}{y}\right)exp\left(-\frac{H^2}{2}\gamma\left(x-\frac{1}{y}\right)^2\right)dxdy$$

$$J_{14} = \int_0^1 \int_0^1 \frac{1}{x^2 y^2}\left(\frac{1}{x}-\frac{1}{y}\right)^2 exp\left(-2K_e\frac{H}{x}\right)exp\left(-2K_h\frac{H}{y}\right)exp\left(-\frac{H^2}{2}\gamma\left(\frac{1}{x}\right.\right.$$
$$\left.\left.-\frac{1}{y}\right)^2\right)dxdy$$

$$J_{15} = \int_0^1 \int_0^1 \frac{1}{x^2 y^2}\left(\frac{1}{x}-\frac{1}{y}\right)^2 exp\left(-2K_e\frac{H}{x}\right)exp\left(-2K_h\frac{H}{y}\right)exp\left(-\frac{H^2}{2}\gamma\left(\frac{1}{x}\right.\right.$$
$$\left.\left.-\frac{1}{y}\right)^2\right)dxdy$$

$$J_{16} = \int_0^1 \int_0^1 \frac{1}{x^2 y^2}\left(\frac{1}{x}\right.$$
$$\left.+\frac{1}{y}\right)^2 exp\left(-2K_e\frac{H}{x}\right)exp\left(-2K_h\frac{H}{y}\right)exp\left(-\frac{H^2}{2}\gamma\left(\frac{1}{x}+\frac{1}{y}\right)^2\right)d\,xdy$$

$$J_{17} = \int_0^1 \int_0^1 \frac{1}{x^2 y^2}\left(\frac{1}{x}+\frac{1}{y}\right)^2 exp\left(-2K_e\frac{H}{x}\right)exp\left(-2K_h\frac{H}{y}\right)exp\left(-\frac{H^2}{2}\gamma\left(\frac{1}{x}\right.\right.$$
$$\left.\left.+\frac{1}{y}\right)^2\right)dxdy$$

$$J_{18} = \int_{-1}^1 \int_{-1}^1 (x+y)^2\cos^2\left(\pi_e\frac{x}{2}\right)\cos^2\left(\pi_h\frac{y}{2}\right)exp\left(-\frac{H^2}{2}\gamma(x+y)^2\right)dxdy$$

$$Z_3 = -\frac{H^3}{16}\left(K_e B_e^2 J_{10} + K_e B_e^2 J_{11} + K_e B_e^2 B_h^2 J_{14} + K_e B_e^2 B_h^2 J_{15} + K_e B_e^2 B_h^2 J_{16}\right.$$
$$\left.+ K_e B_e^2 B_h^2 J_{17} - \frac{\pi_e}{H}B_h^2 J_{19} - \frac{\pi_e}{H}B_h^2 J_{20} - \frac{\pi_e}{H}J_{21}\right)$$

$$J_{19} = \int_{-1}^1 \int_0^1 \frac{1}{x^2}\left(\frac{1}{x}\right.$$
$$\left.+y\right)^2 exp\left(-2K_h\frac{H}{x}\right)\sin\left(\pi_e\frac{y}{2}\right)\cos\left(\pi_e\frac{y}{2}\right)exp\left(-\frac{H^2}{2}\gamma\left(\frac{1}{x}\right.\right.$$
$$\left.\left.+y\right)^2\right)dxdy$$

$$J_{20} = \int_{-1}^{1} \int_{0}^{1} \frac{1}{x^2} \Big(\frac{1}{x}$$
$$-y \Big)^2 exp\left(-2K_h \frac{H}{x}\right) \sin\left(\pi_e \frac{y}{2}\right) \cos\left(\pi_e \frac{y}{2}\right) exp\left(-\frac{H^2}{2}\gamma\left(\frac{1}{x}\right.\right.$$
$$\left.\left.-y\right)^2\right) dxdy$$

$$J_{21} = \int_{-1}^{1} \int_{-1}^{1} (x+y)^2 \cos^2\left(\pi_h \frac{x}{2}\right) \sin\left(\pi_e \frac{y}{2}\right) \cos\left(\pi_e \frac{y}{2}\right) exp\left(-\frac{H^2}{2}\gamma(x+y)^2\right) dxdy$$

$$Z_4 = -\frac{H^3}{8}\Big(K_h B_h^2 J_{12} + K_h B_h^2 J_{13} + K_h B_e^2 B_h^2 J_{14} + K_h B_e^2 B_h^2 J_{15} + K_h B_e^2 B_h^2 J_{16}$$
$$+ K_h B_e^2 B_h^2 J_{17} - \frac{\pi_h}{H} B_e^2 J_{22} - \frac{\pi_h}{H} B_e^2 J_{23} - \frac{\pi_h}{H} J_{24}\Big)$$

$$J_{22} = \int_{0}^{1} \int_{-1}^{1} \frac{1}{x^2} \Big(\frac{1}{x}$$
$$+y \Big)^2 exp\left(-2K_e \frac{H}{x}\right) \sin\left(\pi_h \frac{y}{2}\right) \cos\left(\pi_h \frac{y}{2}\right) exp\left(-\frac{H^2}{2}\gamma\left(\frac{1}{x}\right.\right.$$
$$\left.\left.+y\right)^2\right) dxdy$$

$$J_{23} = \int_{0}^{1} \int_{-1}^{1} \frac{1}{x^2} \Big(\frac{1}{x}$$
$$-y \Big)^2 exp\left(-2K_e \frac{H}{x}\right) \sin\left(\pi_h \frac{y}{2}\right) \cos\left(\pi_h \frac{y}{2}\right) exp\left(-\frac{H^2}{2}\gamma\left(\frac{1}{x}\right.\right.$$
$$\left.\left.-y\right)^2\right) dxdy$$

$$J_{24} = \int_{-1}^{1} \int_{-1}^{1} (x+y)^2 \cos^2\left(\pi_e \frac{x}{2}\right) \sin\left(\pi_h \frac{y}{2}\right) \cos\left(\pi_h \frac{y}{2}\right) exp\left(-\frac{H^2}{2}\gamma(x+y)^2\right) dxdy$$

Integrals from the Coulomb potential energy operator . $\langle V_{coul} \rangle$

$$Z_c = \frac{H^2}{4}(B_e^2 J_{25} + B_e^2 J_{26} + B_h^2 J_{27} + B_h^2 J_{28} + B_e^2 B_h^2 J_{29} + B_e^2 B_h^2 J_{30} + B_e^2 B_h^2 J_{31}$$
$$+ B_e^2 B_h^2 J_{32} + J_{33})$$

$$J_{25} = \int_{0}^{1} \int_{-1}^{1} \frac{1}{x^2} exp\left(-2K_e \frac{H}{x}\right) \cos^2\left(\pi_h \frac{y}{2}\right) \frac{exp\left(-\frac{H^2}{2}\gamma\left(\frac{1}{x}+y\right)^2\right)}{\frac{P_1(\alpha,R)}{P_3(\alpha,R)} + \frac{H}{2}\left|\frac{1}{x}+y\right|} dxdy$$

$$J_{26} = \int_{0}^{1} \int_{-1}^{1} \frac{1}{x^2} exp\left(-2K_e \frac{H}{x}\right) \cos^2\left(\pi_h \frac{y}{2}\right) \frac{exp\left(-\frac{H^2}{2}\gamma\left(\frac{1}{x}-y\right)^2\right)}{\frac{P_1(\alpha,R)}{P_3(\alpha,R)} + \frac{H}{2}\left|\frac{1}{x}-y\right|} dxdy$$

$$J_{27} = \int_{-1}^{1} \int_{0}^{1} \frac{1}{y^2} \cos^2\left(\pi_e \frac{x}{2}\right) exp\left(-2K_h \frac{H}{y}\right) \frac{exp\left(-\frac{H^2}{2}\gamma\left(x + \frac{1}{y}\right)^2\right)}{\frac{P_1(\alpha,R)}{P_3(\alpha,R)} + \frac{H}{2}\left|x + \frac{1}{y}\right|} \, dxdy$$

$$J_{28} = \int_{-1}^{1} \int_{0}^{1} \frac{1}{y^2} \cos^2\left(\pi_e \frac{x}{2}\right) exp\left(-2K_h \frac{H}{y}\right) \frac{exp\left(-\frac{H^2}{2}\gamma\left(x - \frac{1}{y}\right)^2\right)}{\frac{P_1(\alpha,R)}{P_3(\alpha,R)} + \frac{H}{2}\left|x - \frac{1}{y}\right|} \, dxdy$$

$$J_{29} = \int_{0}^{1} \int_{0}^{1} \frac{1}{x^2}\frac{1}{y^2} exp\left(-2K_e \frac{H}{x}\right) exp\left(-2K_h \frac{H}{y}\right) \frac{exp\left(-\frac{H^2}{2}\gamma\left(\frac{1}{x} - \frac{1}{y}\right)^2\right)}{\frac{P_1(\alpha,R)}{P_3(\alpha,R)} + \frac{H}{2}\left|\frac{1}{x} - \frac{1}{y}\right|} \, dxdy$$

$$J_{30} = \int_{0}^{1} \int_{0}^{1} \frac{1}{x^2}\frac{1}{y^2} exp\left(-2K_e \frac{H}{x}\right) exp\left(-2K_h \frac{H}{y}\right) \frac{exp\left(-\frac{H^2}{2}\gamma\left(\frac{1}{x} - \frac{1}{y}\right)^2\right)}{\frac{P_1(\alpha,R)}{P_3(\alpha,R)} + \frac{H}{2}\left|\frac{1}{x} - \frac{1}{y}\right|} \, dxdy$$

$$J_{31} = \int_{0}^{1} \int_{0}^{1} \frac{1}{x^2}\frac{1}{y^2} exp\left(-2K_e \frac{H}{x}\right) exp\left(-2K_h \frac{H}{y}\right) \frac{exp\left(-\frac{H^2}{2}\gamma\left(\frac{1}{x} + \frac{1}{y}\right)^2\right)}{\frac{P_1(\alpha,R)}{P_3(\alpha,R)} + \frac{H}{2}\left|\frac{1}{x} + \frac{1}{y}\right|} \, dxdy$$

$$J_{32} = \int_{0}^{1} \int_{0}^{1} \frac{1}{x^2}\frac{1}{y^2} exp\left(-2K_e \frac{H}{x}\right) exp\left(-2K_h \frac{H}{y}\right) \frac{exp\left(-\frac{H^2}{2}\gamma\left(\frac{1}{x} + \frac{1}{y}\right)^2\right)}{\frac{P_1(\alpha,R)}{P_3(\alpha,R)} + \frac{H}{2}\left|\frac{1}{x} + \frac{1}{y}\right|} \, dxdy$$

$$J_{33} = \int_{-1}^{1} \int_{-1}^{1} \cos^2\left(\pi_e \frac{x}{2}\right) \cos^2\left(\pi_h \frac{y}{2}\right) \frac{exp\left(-\frac{H^2}{2}\gamma(x + y)^2\right)}{\frac{P_1(\alpha,R)}{P_3(\alpha,R)} + \frac{H}{2}|x + y|} \, dxdy$$

Appendix B

$$\left\langle \Psi^* \left| -\left(\frac{d^2}{dr^2} + \frac{1}{r}\frac{d}{dr}\right) \right| \Psi \right\rangle = -\int_0^{+\infty} \Psi^* \left(\frac{d^2}{dr^2} + \frac{1}{r}\frac{d}{dr}\right) \Psi \, r \, dr$$

$$I = -\int_0^{+\infty} \Psi^* \left(\frac{d^2}{dr^2} + \frac{1}{r}\frac{d}{dr}\right) \Psi \, r \, dr = -\int_0^{+\infty} \Psi^* \frac{1}{r}\frac{d}{dr}\left(r\frac{d}{dr}\right) \Psi \, r \, dr$$

With Ψ is the eigenfunction

$$\int u \, dv = uv - \int v \, du$$

Let $u = \Psi^* \Rightarrow du = \frac{d\,\Psi^*}{dr} dr$ and $dv = \frac{d}{dr}\left(r\frac{d\,\Psi}{dr}\right) dr \;\; v = (r\frac{d\,\Psi}{dr})$

Hence

$$I = -\left[\Psi^* r \frac{d\,\Psi}{dr}\right]_0^{\infty} + \int_0^{+\infty} r \frac{d\Psi}{dr}\frac{d\,\Psi^*}{dr} dr$$

But $\left[\Psi^* r \frac{d\,\Psi}{dr}\right]_0^{\infty} = 0$ because $\Psi(\infty) = 0$

Ψ Is a real function $\Psi^* \Psi = \Psi^2, \frac{d\Psi}{dr}\frac{d\,\Psi^*}{dr} \Longrightarrow \left(\frac{d\Psi}{dr}\right)^2$

This means we can rewrite the equation

$$I = \int_0^{+\infty} r \frac{d\Psi}{dr}\frac{d\,\Psi^*}{dr} dr = \int_0^{+\infty} \left(\frac{d\Psi}{dr}\right)^2 dr$$

Integrals P_i is defined by :

$P_1 = \int_0^{+\infty} \int_0^{+\infty} \int_0^{+\infty} (f(r_{e1}) f(r_{e2}) f(r_h) \aleph)^2 \, r_{e1} dr_{e1} \, r_{e2} dr_{e2} r_h dr_h$

$P_2 = \int_0^{+\infty} \int_0^{+\infty} \int_0^{+\infty} (f(r_{e1}) f(r_{e2}) f(r_h) \frac{d\aleph}{dr_{e1}})^2 \, r_{e1} dr_{e1} \, r_{e2} dr_{e2} r_h dr_h$

$P_3 = \int_0^{+\infty} \int_0^{+\infty} \int_0^{+\infty} \frac{1}{|r_{e1}-r_{e2}|} (f(r_{e1}) f(r_{e2}) f(r_h) \aleph)^2 \, r_{e1} dr_{e1} \, r_{e2} dr_{e2} r_h dr_h$

$P_4 = \int_0^{+\infty} \int_0^{+\infty} \int_0^{+\infty} (f(r_{e2}) f(r_h))^2 f(r_{e1}) f'(r_{e1}) \aleph \frac{d\aleph}{dr_{e1}} r_{e1} dr_{e1} \, r_{e2} dr_{e2} r_h dr_h$

$P_6 = \int_0^{+\infty} \int_0^{+\infty} \int_0^{+\infty} (f'(r_{e1}) f(r_{e2}) f(r_h) \aleph)^2 \, r_{e1} dr_{e1} \, r_{e2} dr_{e2} r_h dr_h$

$P_7 = \int_0^{R} \int_0^{+\infty} \int_0^{+\infty} (f(r_{e1}) f(r_{e2}) f(r_h) \aleph)^2 \, r_{e1} dr_{e1} \, r_{e2} dr_{e2} r_h dr_h$

$P_{2A} = \int_0^{+\infty} \int_0^{+\infty} \int_0^{+\infty} (f(r_{e1}) f(r_{e2}) f(r_h) \frac{d\aleph}{dr_{e2}})^2 \, r_{e1} dr_{e1} \, r_{e2} dr_{e2} r_h dr_h$

$$P_{3A}=\int_0^{+\infty}\int_0^{+\infty}\int_0^{+\infty}\frac{1}{|r_{e1}-r_h|}(f(r_{e1})\,f(r_{e2})f(r_h)\aleph)^2\,r_{e1}dr_{e1}\,r_{e2}dr_{e2}r_hdr_h$$

$$P_{4A}=\int_0^{+\infty}\int_0^{+\infty}\int_0^{+\infty}(f(r_{e1})f(r_h))^2f(r_{e2})f'(r_{e2})\aleph\frac{d\aleph}{dr_{e2}}r_{e1}dr_{e1}\,r_{e2}dr_{e2}r_hdr_h$$

$$P_{6A}=\int_0^{+\infty}\int_0^{+\infty}\int_0^{+\infty}(f(r_{e1})\,f'(r_{e2})f(r_h)\aleph)^2\,r_{e1}dr_{e1}\,r_{e2}dr_{e2}r_hdr_h$$

$$P_{7A}=\int_0^{+\infty}\int_0^{R}\int_0^{+\infty}(f(r_{e1})\,f(r_{e2})f(r_h)\aleph)^2\,r_{e1}dr_{e1}\,r_{e2}dr_{e2}r_hdr_h$$

$$P_{2B}=\int_0^{+\infty}\int_0^{+\infty}\int_0^{+\infty}(f(r_{e1})\,f(r_{e2})f(r_h)\frac{d\aleph}{dr_h})^2\,r_{e1}dr_{e1}\,r_{e2}dr_{e2}r_hdr_h$$

$$P_{3B}=\int_0^{+\infty}\int_0^{+\infty}\int_0^{+\infty}\frac{1}{|r_{e2}-r_h|}(f(r_{e1})\,f(r_{e2})f(r_h)\aleph)^2\,r_{e1}dr_{e1}\,r_{e2}dr_{e2}r_hdr_h$$

$$P_{4B}=\int_0^{+\infty}\int_0^{+\infty}\int_0^{+\infty}(f(r_{e1})f(r_{e2}))^2f(r_h)f'(r_h)\aleph\frac{d\aleph}{dr_h}r_{e1}dr_{e1}\,r_{e2}dr_{e2}r_hdr_h$$

$$P_{6B}=\int_0^{+\infty}\int_0^{+\infty}\int_0^{+\infty}(f(r_{e1})\,f(r_{e2})f'(r_h)\aleph)^2\,r_{e1}dr_{e1}\,r_{e2}dr_{e2}r_hdr_h$$

$$P_{7B}=\int_0^{+\infty}\int_0^{+\infty}\int_0^{R}(f(r_{e1})\,f(r_{e2})f(r_h)\aleph)^2\,r_{e1}dr_{e1}\,r_{e2}dr_{e2}r_hdr_h$$

With $\aleph$ is the correlation function of the radial part with expression : $\aleph = exp[(-\alpha_1|r_{e1}-r_{e2}|-\alpha_2|r_{e1}-r_h|-\alpha_3|r_{e2}-r_h|)]$

The integral z_i is defined by :

$$Z_1=\int_{-\infty}^{+\infty}\int_{-\infty}^{+\infty}\int_{-\infty}^{+\infty}(g(z_{e1})\,g(z_{e2})g(z_h)\zeta)^2\,dz_{e1}\,dz_{e2}dz_h$$

$$Z_2=\int_{-\infty}^{+\infty}\int_{-\infty}^{+\infty}\int_{-\infty}^{+\infty}(g(z_{e1})\,g(z_{e2})g(z_h))^2\left(\frac{d\zeta}{dz_{e1}}\right)^2\,dz_{e1}\,dz_{e2}dz_h$$

$$Z_3=\int_{-\infty}^{+\infty}\int_{-\infty}^{+\infty}\int_{-\infty}^{+\infty}(g(z_{e2})g(z_h))^2g(z_{e1})g'(z_{e1})\zeta\left(\frac{d\zeta}{dz_{e1}}\right)\,dz_{e1}\,dz_{e2}dz_h$$

$$Z_5=\int_{-\frac{H}{2}}^{+\frac{H}{2}}\int_{-\infty}^{+\infty}\int_{-\infty}^{+\infty}(g(z_{e1})\,g(z_{e2})g(z_h)\zeta)^2\,dz_{e1}\,dz_{e2}dz_h$$

$$Z_{2A}=\int_{-\infty}^{+\infty}\int_{-\infty}^{+\infty}\int_{-\infty}^{+\infty}(g(z_{e1})\,g(z_{e2})g(z_h))^2\left(\frac{d\zeta}{dz_{e2}}\right)^2\,dz_{e1}\,dz_{e2}dz_h$$

$$Z_{3A}=\int_{-\infty}^{+\infty}\int_{-\infty}^{+\infty}\int_{-\infty}^{+\infty}(g(z_{e1})g(z_h))^2g(z_{e2})g'(z_{e2})\zeta\left(\frac{d\zeta}{dz_{e2}}\right)\,dz_{e1}\,dz_{e2}dz_h$$

$$Z_{5A}=\int_{-\infty}^{+\infty}\int_{-\frac{H}{2}}^{+\frac{H}{2}}\int_{-\infty}^{+\infty}(g(z_{e1})\,g(z_{e2})g(z_h)\zeta)^2\,dz_{e1}\,dz_{e2}dz_h$$

$$Z_{2B}=\int_{-\infty}^{+\infty}\int_{-\infty}^{+\infty}\int_{-\infty}^{+\infty}(g(z_{e1})\,g(z_{e2})g(z_h))^2\left(\frac{d\zeta}{dz_h}\right)^2\,dz_{e1}\,dz_{e2}dz_h$$

$$Z_{3B}=\int_{-\infty}^{+\infty}\int_{-\infty}^{+\infty}\int_{-\infty}^{+\infty}(g(z_{e1})g(z_{e2}))^2g(z_h)g'(z_h)\zeta\left(\frac{d\zeta}{dz_h}\right)\,dz_{e1}\,dz_{e2}dz_h$$

$$Z_{5B}=\int_{-\infty}^{+\infty}\int_{-\infty}^{+\infty}\int_{-\frac{H}{2}}^{+\frac{H}{2}}(g(z_{e1})\,g(z_{e2})g(z_h)\zeta)^2\,dz_{e1}\,dz_{e2}dz_h$$

$$Z_C=\int_{-\infty}^{+\infty}\int_{-\infty}^{+\infty}\int_{-\infty}^{+\infty}\frac{1}{\frac{P_1}{P_3}+|z_{e1}-z_{e2}|}(g(z_{e1})\,g(z_{e2})g(z_h)\zeta)^2\,dz_{e1}\,dz_{e2}dz_h$$

$$Z_{CA}=\int_{-\infty}^{+\infty}\int_{-\infty}^{+\infty}\int_{-\infty}^{+\infty}\frac{1}{\frac{P_1}{P_{3A}}+|z_{e1}-z_h|}(g(z_{e1})\,g(z_{e2})g(z_h)\zeta)^2\,dz_{e1}\,dz_{e2}dz_h$$

$$Z_{CB}=\int_{-\infty}^{+\infty} \int_{-\infty}^{+\infty} \int_{-\infty}^{+\infty} \frac{1}{\frac{P_1}{P_{3B}}+|z_{e2}-z_h|} \left(g(z_{e1})\,g(z_{e2})g(z_h)\zeta\right)^2 dz_{e1}\,dz_{e2}dz_h$$

Withζ is the correlation function of the axial part with expression :

$$\zeta = exp[(-\gamma_1|z_{e1}-z_{e2}| - \gamma_2|z_{e1}-z_h| - \gamma_3|z_{e2}-z_h|)]$$

Appendix C

Numerical calculation flowchart

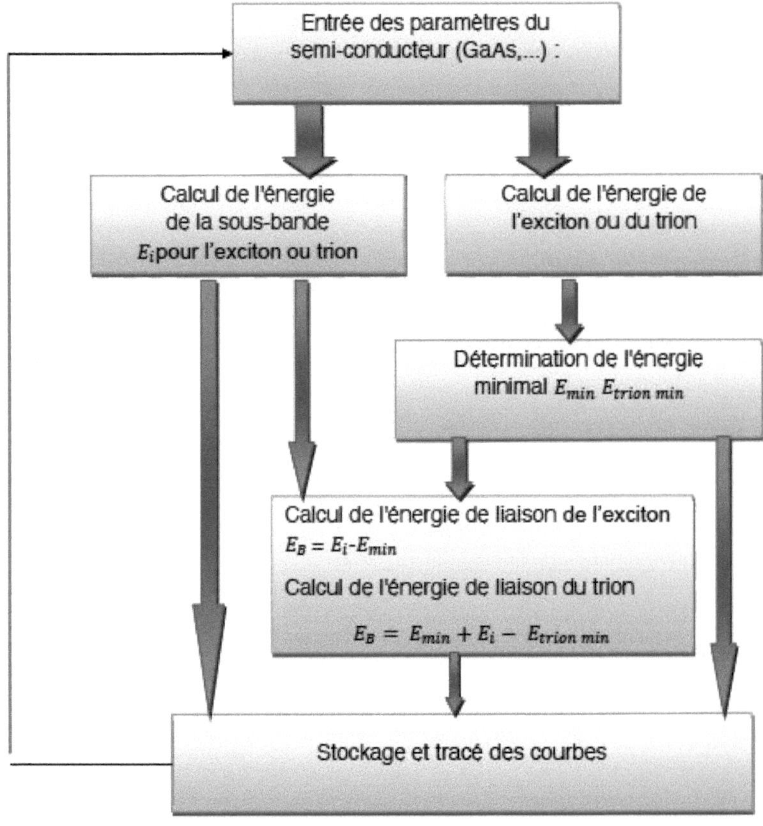

Printed by Books on Demand GmbH, Norderstedt / Germany